# LIFE SCIENCE

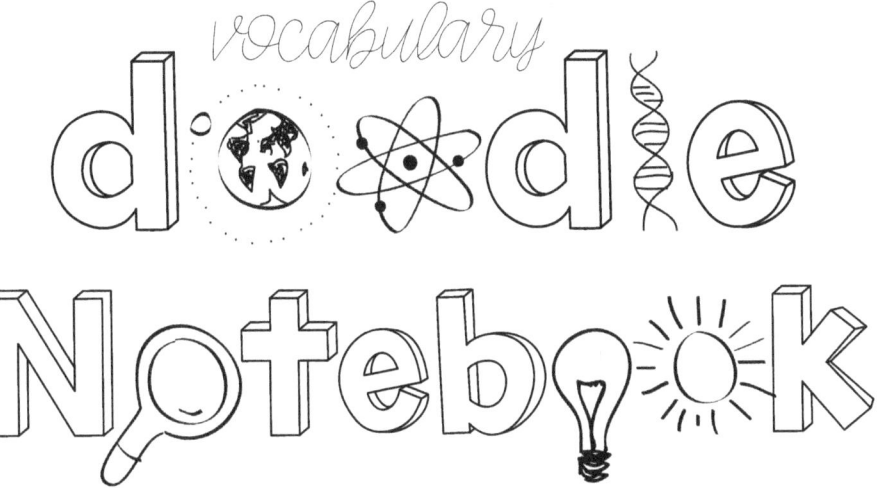

vocabulary doodle Notebook

Thanks so much for purchasing this book!

This resource is licensed to be used by a single student only. The copyright is owned by CAPTIVATE SCIENCE and all rights are reserved.

**Copying pages is prohibited.**

© Copyright 2021 Captivate Science

Thanks and credit to the following artists:

To learn more about the doodle note method and download your free Doodle Note Handbook, visit **doodlenotes.org** and **sciencedoodlenotes.com**

# notebook contents

## Intro to Visual Notes

| | |
|---|---|
| Benefits of the Doodle Note Method | 5 |
| Doodle Notes and Your Brain | 7 |
| Doodle Note Tips | 8 |

## Chapter 1: Molecules to Organisms: Structures and Processes

| | |
|---|---|
| MS LS1-1 Living Thing Are Made of Cells | 11 |
| MS LS1-2 How Cells Function as Whole System | 14 |
| MS LS1-3 The Body System of Interacting Subsystems | 18 |
| MS LS1-4 How Behaviors Impact Reproduction | 38 |
| MS LS1-5 Environmental and Genetic Factors Impact Growth | 42 |
| MS LS1-6 Photosynthesis 101 | 45 |
| MS LS1-7 Matter and Energy in Organisms | 47 |
| MS LS1-8 Stimuli, Memory and Behavior | 49 |

## Chapter 2: Interactions, Energy, and Dynamics Relationships in Ecosystems

| | |
|---|---|
| MS LS2-1 Resources and Populations of Organisms | 53 |
| MS LS2-2 Interactions Between Organisms | 55 |
| MS LS2-3 Matter and Energy in an Ecosystem | 58 |

© 2021 Captivate Science

# notebook contents

## Chapter 2: Interactions, Energy, and Dynamics Relationships in Ecosystems

| | |
|---|---|
| MS LS2-4 Ecosystem Dynamics | 63 |
| MS LS2-5 Maintaining Biodiversity | 65 |

## Chapter 3: Heredity: Inheritance and Variation of Traits

| | |
|---|---|
| MS LS3-1 Changes in Genetic Material | 68 |
| MS LS3-2 Asexual vs. Sexual Reproduction | 70 |

## Chapter 4: Biological Evolution: Unity and Diversity

| | |
|---|---|
| MS LS4-1 Change of Life Throughout History of Earth | 75 |
| MS LS4-2 Inferring Evolutionary Relationships | 78 |
| MS LS4-3 Comparing Embryo Development of Different Organisms | 80 |
| MS LS4-4 Traits and Survival | 81 |
| MS LS4-5 Human Influence on Inheritance of Traits | 82 |
| MS LS4-6 Changes in Populations Over time | 83 |

## Doodle DIY:  Templates for Additional Notes        85

Get doodle help for each page at **captivatescience.com**

# LIFE SCIENCE doodle VOCABULARY NOTES

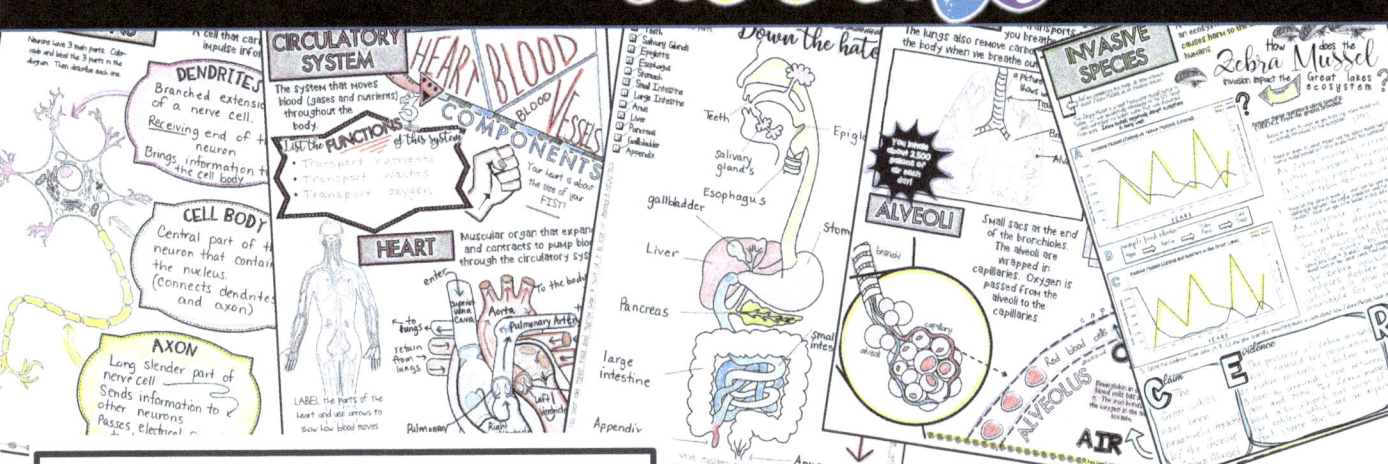

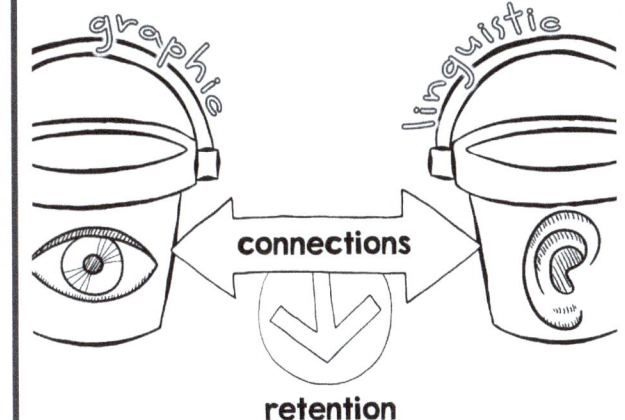

Students can retain more information when they connect the linguistic and visual centers of the brain.

The referential connections between the two zones allow information to be stored and become memory.

## the brain converts

**BLENDS** of *visual* information and *linguistic* information into long-term memory more easily.

*Visual note-taking* allows student brains to process and retain lesson material more effectively.

doodlenotes.org

# What are the learning benefits of visual note-taking?

## doodle notes®

 stronger **focus**

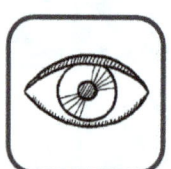 **retention** through dual coding

 **mental** connections

 memory **boost**

 communication between **brain** hemispheres

 **building** long-term memories

 activated **neural** pathways

 increased **creativity** & alertness

 associative recognition

 **picture** superiority **effect**

 relaxation **benefits**

 **problem solving** skills boost

# GETTING STARTED

☐ **Access the Learning Guides** at www.captivatescience.com
Each page in this book comes with a digital learning guide where you can gather information to complete each page of your doodle notes.

☐ Complete the **Doodle Notes and Your Brain** Doodle Page to learn how your brain can benefit from the Doodle Note Method.

☐ Complete the **Doodle Notes Tips** Doodle Page

*How can I use doodle notes to help me learn about life science?*

# Doodle Notes and Your Brain

1. Write/copy in the left-brain activities inside of the left hemisphere of the brain. Write in the right-brain activities in the right hemisphere.

2. Label the corpus callosum.

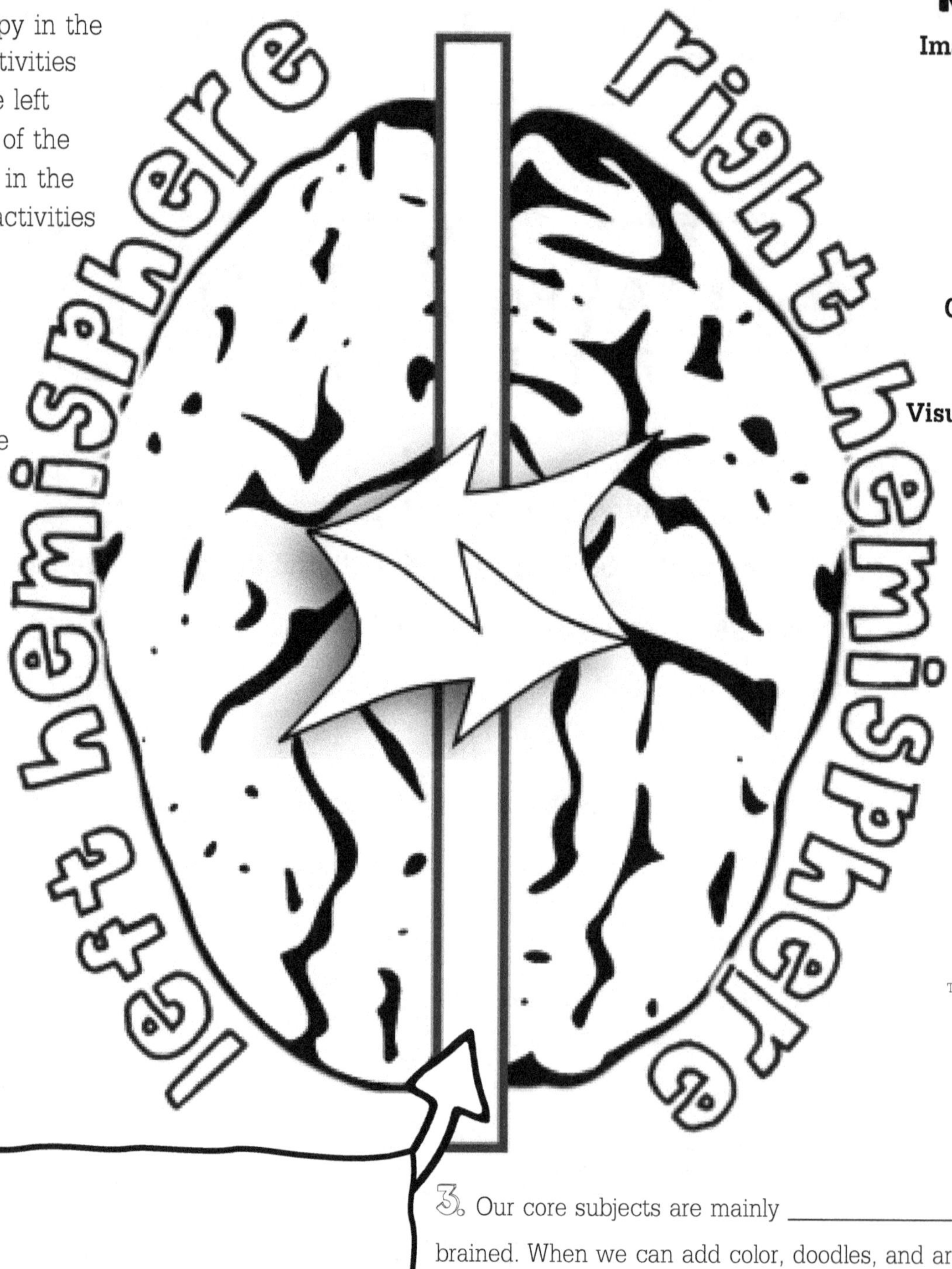

**left hemisphere** **right hemisphere**

**RIGHT**
Imagination
Rhythm
Arts
Color
Music
Creativity
Intuition
Visualization

**LEFT**
Logic
Facts
Sequencing
Language
Analysis
Reading
Writing
Science

This page was created by The Doodle Note Method and mathgiraffe.com. To learn more visit the doodle note headquarters at doodlenotes.org.

3. Our core subjects are mainly _____ brained. When we can add color, doodles, and artistic embellishment, we incorporate the _____ brain.

4. Communication between both sides of the brain at the same time activates our brains more fully to help us to maximize learning: LIST 3 BENEFITS OF DOODLE NOTES IN THE SPACE BELOW

# Doodle Note Tips

Doodle notes take practice. As you complete this book your skills will grow! Maximize your learning with these doodle tips....

Use the size of your letters to emphasize importance. Main idea (LARGE), subtopics (MEDIUM), details (SMALL).

<<<< **Practice writing some words in different sizes here!** Do you notice how the larger words stand out?

Use bullet points and short phrases. This doesn't mean you are cutting corners! Be sure to write down the essentials that will help you learn the information! Have fun with the style of your bullet points. They don't have to just be DOTS! **Practice here >>>>**

Shade each letter using the technique below.

# LOOK!

- Full weight dark shading
- Hatching
- Dotting or dashing
- VERY lightly shading with a dark outline
- zigzagging

Can you think of some OTHER WAYS to highlight key ideas? SHOW YOUR DOODLE SKILLS BELOW:

**Practice here >>>**

There are SO MANY ways to make information stand out in your doodle notes. In the space above, practice some highlighting techniques.

© 2021 Captivate Science

# CHAPTER 1

## Molecules to Organisms: Structures and Processes

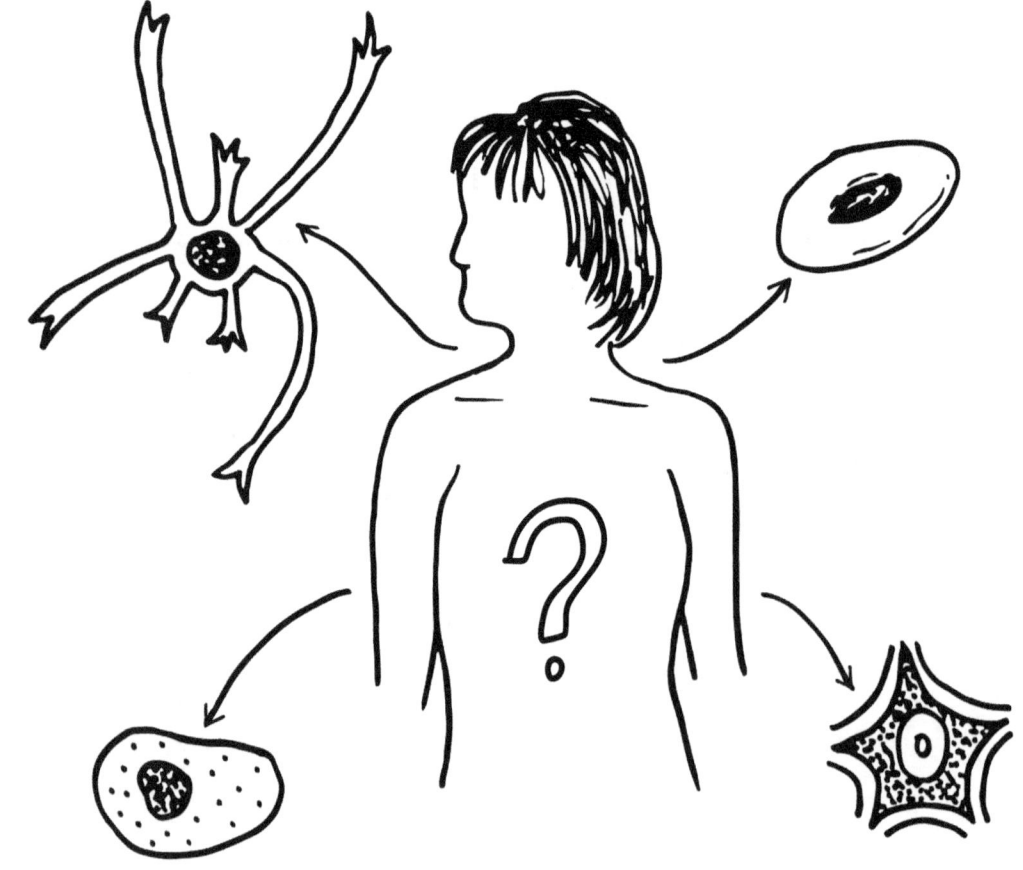

How do cells contribute to the function of living organisms?

MS-LS1-1

# The CELL

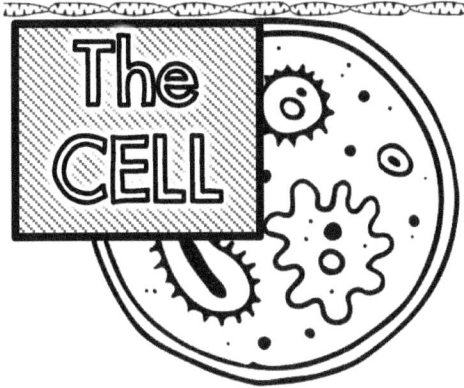

Structural unit of life.

Cells are the smallest **LIVING** units.

Some living things have ONE cell, while other living things consist of MILLIONS of different cells!

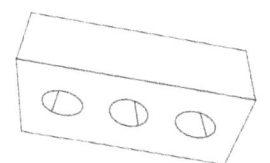

*Living things are made of one or many cells!*

## UNICELLULAR VS. MULTICELLULAR

**The body is made up of only one cell.**

Facts

*Paramecium*  *Amoeba*

**The body is made up MANY cells working together.**

Facts

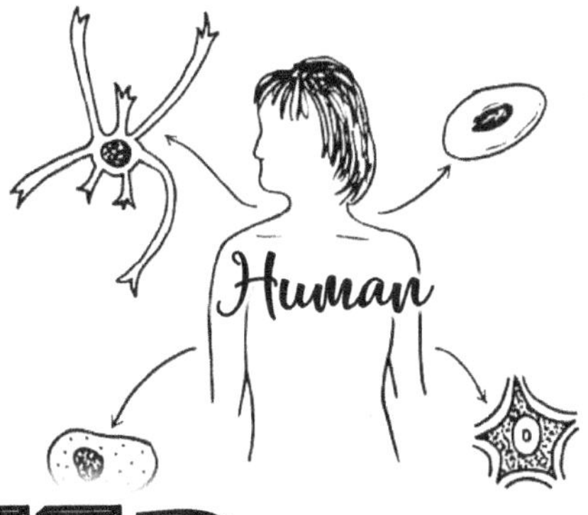

*Human*

## BOTH NEED

*Doodle a picture for each of the things that they both need!*

| FOOD | | WASTE DISPOSAL | ENVIRONMENT |

WATER

# CELL THEORY

MS LS1-1

Research cell theory and add 3 or more events to this timeline.

A collection of conclusions from scientists over time that describes cells and how they operate.

| 1600 | 1700 | 1800 | 1900 |

Fill in the blanks, color the pictures.

**1.** Living things are made from _____ or more _____.

DOODLE a picture to match the statement.

CELLS come from existing CELLS by... *division*

**2.**

Fill in the blanks and DOODLE a picture to match.

**3.** An organism's life _____ occur within cells. A cell is the _____ unit in a living thing.

MS LS1-2

A special structure that is responsible for a particular function within a cell. Found only in eukaryotic cells

Draw colorful **doodle arrows** pointing to the cell organelles!

Name some EXAMPLES of cell organelles:

_____

_____

_____

_____

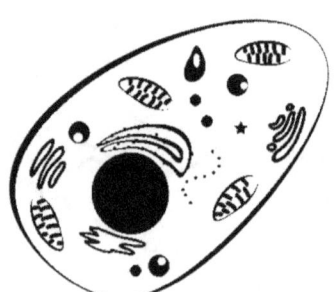

A boundary that controls what enters/exits the cell AND protects cells from the outside environment. **ALL CELLS** have a cell membrane!

DRAW and LABEL a picture of a cell's protective membrane!

Cell membranes are selectively permeable. This means that they select what can pass through and what cannot pass through!

# PLANT CELL

MS LS1-2

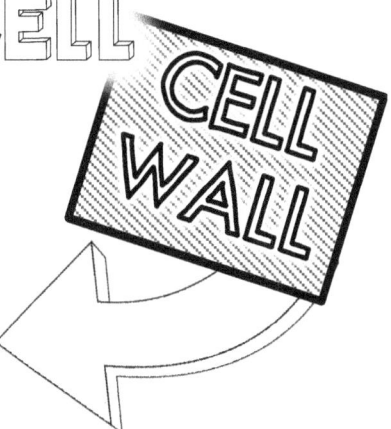

## CELL WALL

The protective outer layer of a *bacteria, fungi or*

It is the outermost layer of a plant cell and is found OUTSIDE the cell membrane.

DRAW a CLOSE UP doodle to show the location of the cell wall compared to the cell membrane!

**How is the cell wall different than a cell membrane?**

## CHLOROPLAST

A specialized organelle found ONLY IN PLANTS.

Chloroplast contain a green pigment called Chlorophyll and have the important job of converting the SUN's energy into CHEMICAL energy via the process of PHOTOSYNTHESIS.

Doodle a picture to match each caption.

A leaf is made of many plant cells. Plant cells consist of different organelles. One plant organelle is called the Chloroplast.

Within one plant cell there are many Chloroplasts working together to make food energy for the cell.

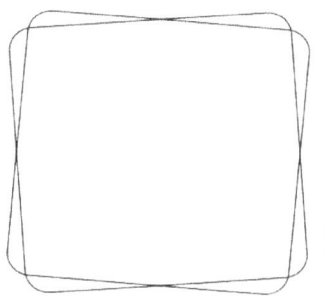

Chloroplasts contain Chlorophyll molecules, which give plants their green color.

The Chloroplasts work to absorb sunlight, water and carbon dioxide, turning it into sugar and oxygen.

MS LS1-2

## STRUCTURE

The arrangement between the elements or parts of something.

*For example: The parts of skeletal muscles.*

THINK AND WRITE: What does this statement mean?

"Structure determines function."
Explain in your own words.

_____
_____
_____
_____
_____
_____

## ANATOMY

The structure of parts in a body.

**Doodle a memory trigger!**
(A memory trigger is a picture to help you remember this definition.)

## FUNCTION

The way something works or operates in a particular way.

*For example: The way muscles contract.*

Can you think of more examples comparing structure and function?

| STRUCTURE | FUNCTION |
|---|---|
| Plants have chloroplasts. → | Plants carry out photosynthesis. |
|  |  |
|  |  |
|  |  |

## PHYSIOLOGY

The functions and relationships of body parts.

**Doodle a memory trigger!**
(A memory trigger is a picture to help you remember this definition.)

# Cells work together to form systems of tissues and organs.

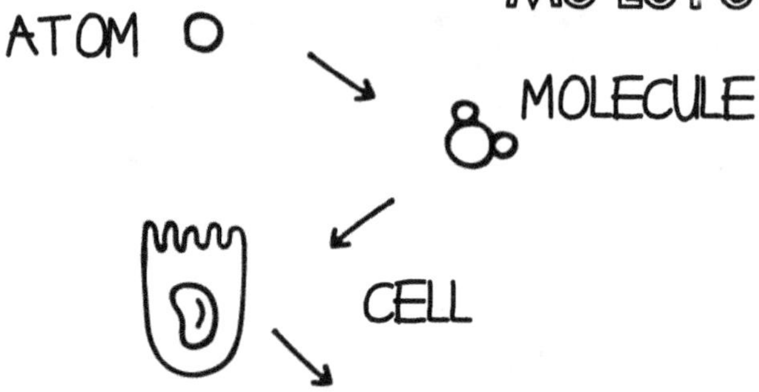

MS LS1-3

ATOM → MOLECULE → CELL

**TISSUE** — Groups of cells that work together for specific functions.

Tissues are grouped together in the body to form organs. These include the brain, heart, lungs, kidneys, liver and more!

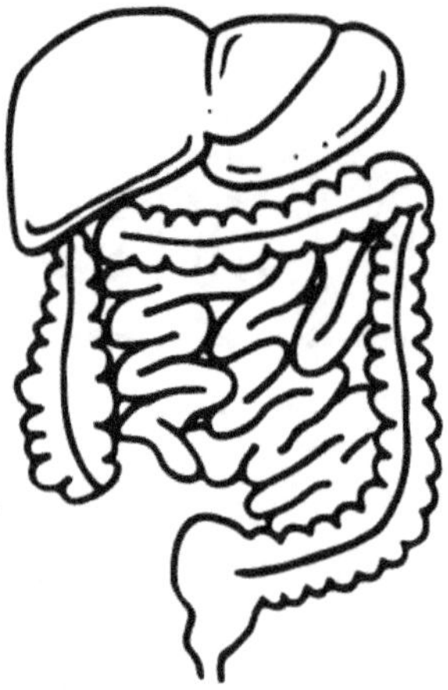

**ORGAN**

**ORGAN SYSTEM** — Different organs working together to perform a common function.

Pick 3 organs that work together to perform a common body function

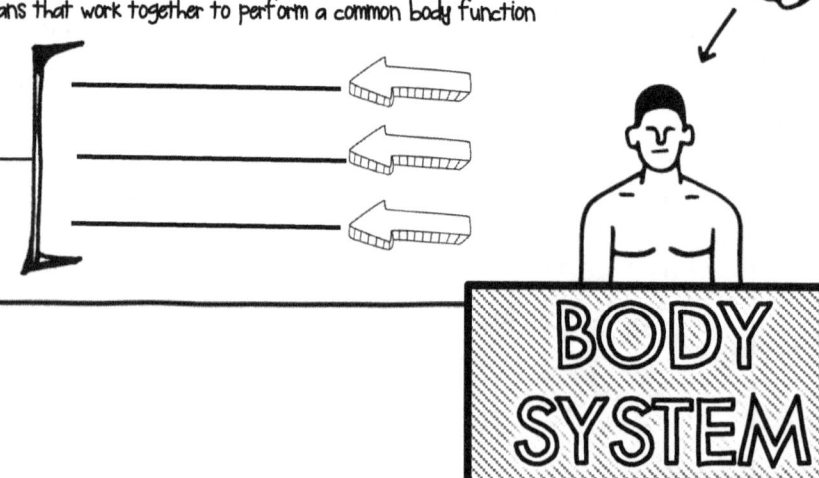

**BODY SYSTEM** — The body is a system of interacting subsystems.

How many body systems and subsystems can you name? Fill the box to the left with a WORDLE of the body systems you can name. Use fun FONTS and Colors!

# CIRCULATORY SYSTEM

MS-LS1-3

The system that moves blood (gases and nutrients) throughout the body.

## List the FUNCTIONS of this system

- 
- 
- 

## 3 COMPONENTS

Your heart is about the size of your FIST!

## HEART

Muscular organ that expands and contracts to pump blood through the circulatory system.

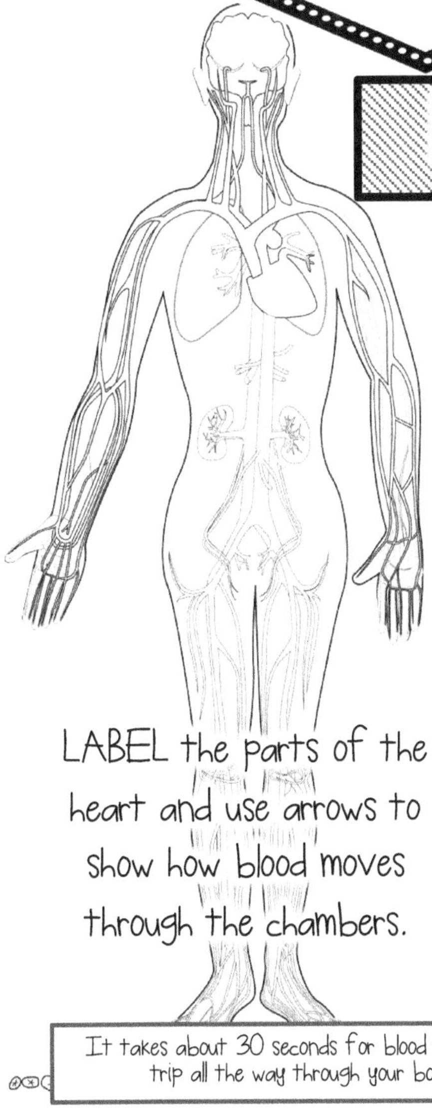

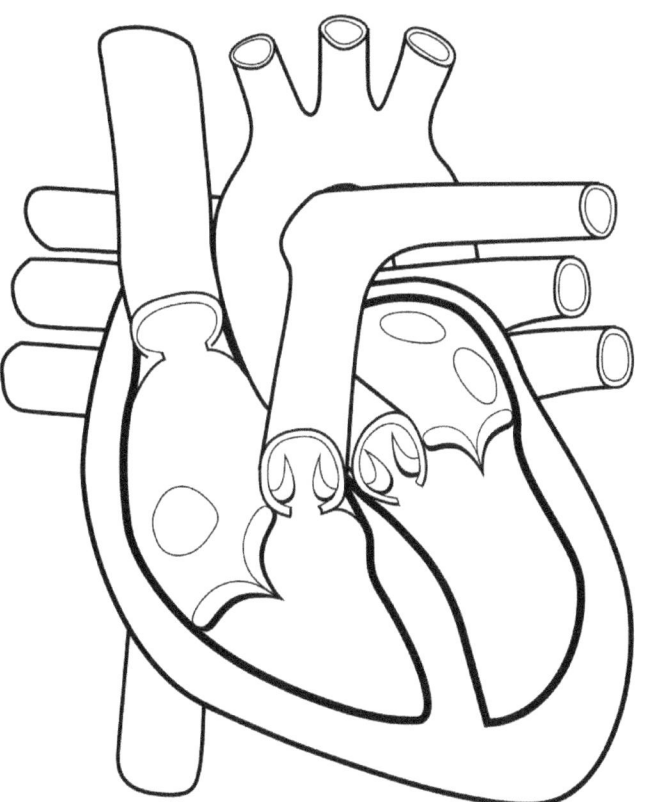

LABEL the parts of the heart and use arrows to show how blood moves through the chambers.

It takes about 30 seconds for blood to make a trip all the way through your body!

© 2021 Captivate Science

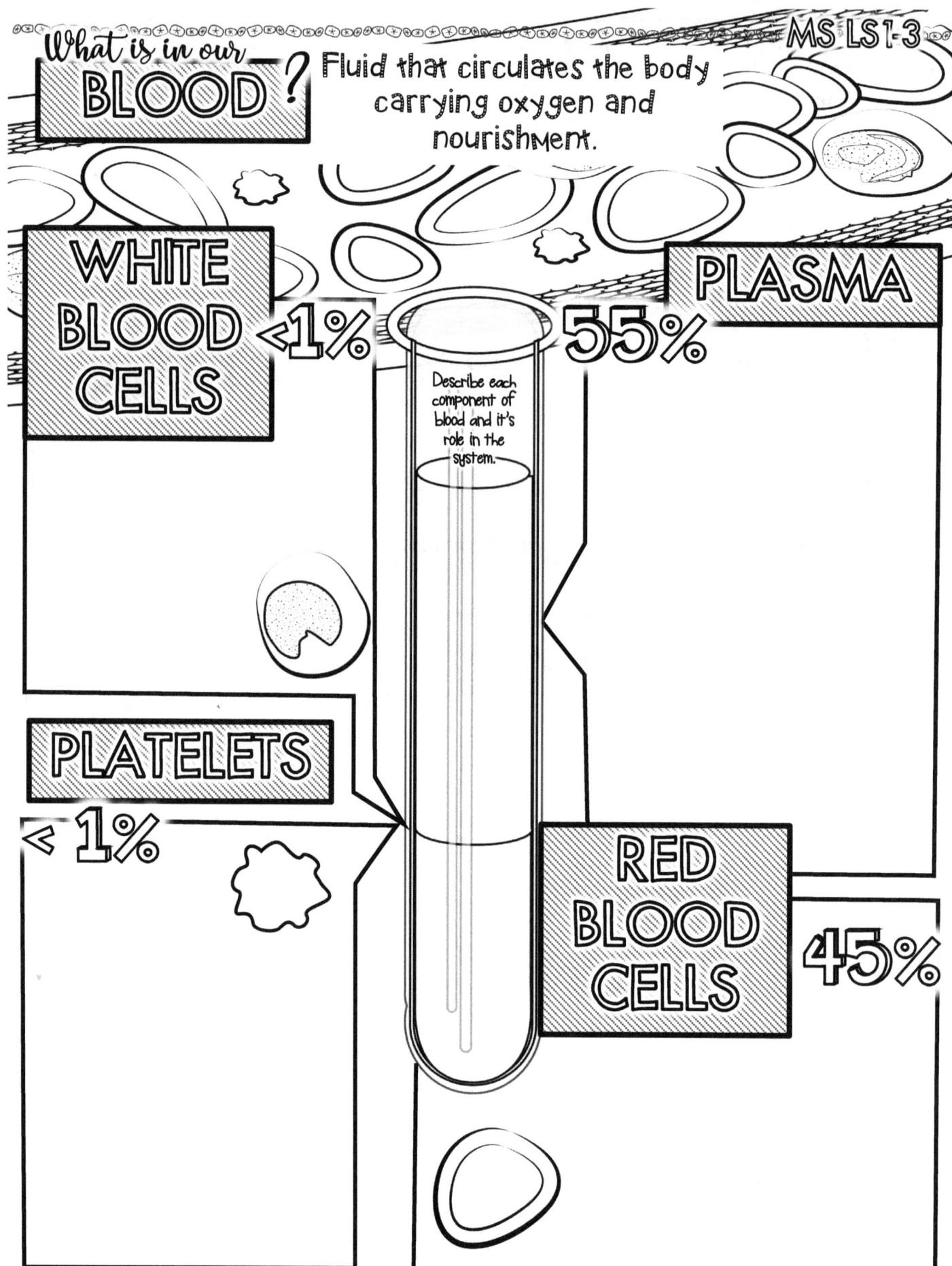

# BLOOD VESSELS

MS-LS1-3

Pathways that carry blood throughout the body.

COLOR and label the blood vessels below! (Artery RED, Capillaries PURPLE and Vein BLUE.)

## ARTERIES

Vessels that carry blood AWAY from the heart to the rest of the body.

**A = AWAY**

## CAPILLARIES

**C = CONNECT**

Vessels that CONNECT arteries to veins. Capillaries are the smallest blood vessels. They are one cell thick. Oxygen, nutrients and waste move easily through capillary walls.

## VEINS

Vessels that bring de-oxygenated blood back to the heart. These LOOK <u>BLUE</u> through our skin!

**B = BACK to the Heart**

Look at the inside of your arm. **Blood is always red,** so why do veins look blue through your skin?

© 2021 Captivate Science

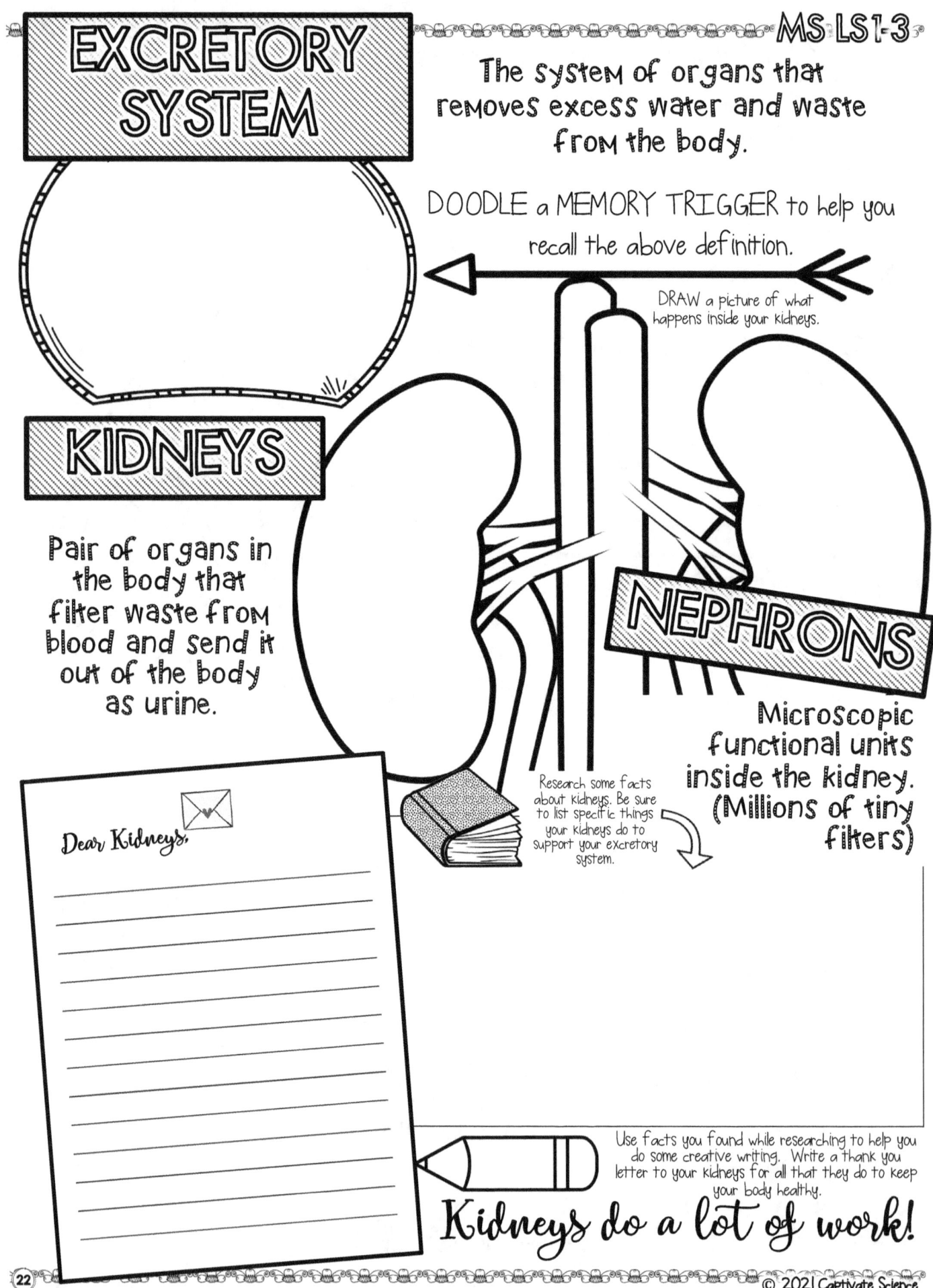

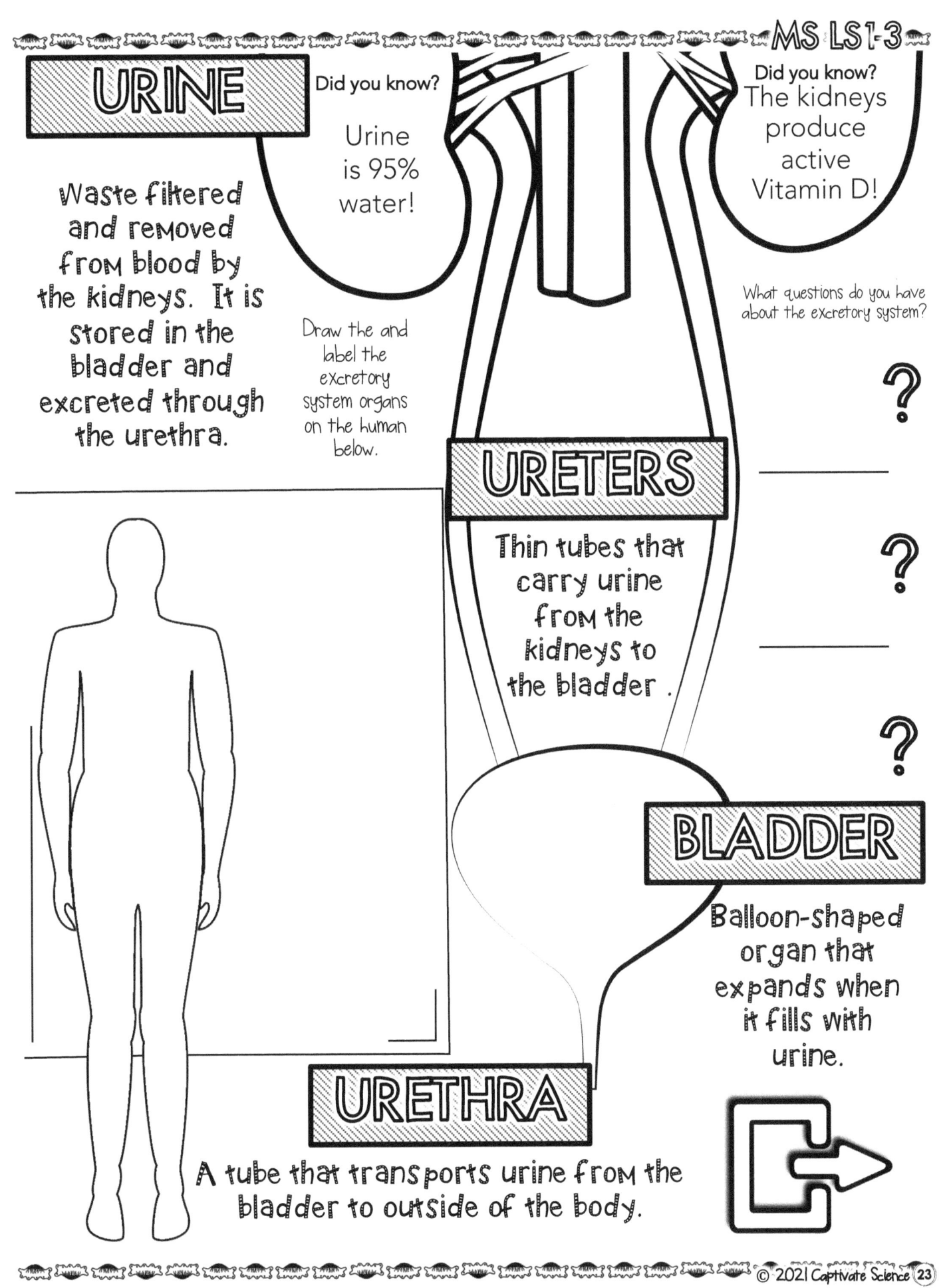

MS LS1-3

# DIGESTIVE SYSTEM

A group of organs that break down food so that it can be used for energy, cell growth and repair. The system starts at the mouth and ends at the anus where things that cannot be digested are pushed out as waste.

You are what you eat!

INGEST
DIGEST
ABSORB
ELIMINATE

## MOUTH

Digestion starts when your teeth break food into smaller pieces that you can swallow.

Describe or doodle 3 different types of teeth in your mouth and their job in breaking down food.

A liquid produced by salivary glands that wets food and contains chemicals that begin to break it down.

## SALIVA

## PERISTALSIS

Pattern of muscle contractions that move food through the esophagus and intestines.

Add these labels to the diagram to the left.
Muscles Contract
Food
Muscles Relax
Movement of Food

◆ Because of peristalsis, food can get to your stomach even if you eat UPSIDE DOWN!!

© 2021 Captivate Science

MS LS1-3

## Down the hatch!

Color and label the digestive system!
Be sure to include these parts:
- ☐ Teeth
- ☐ Salivary Glands
- ☐ Epiglottis
- ☐ Esophagus
- ☐ Stomach
- ☐ Small Intestine
- ☐ Large Intestine
- ☐ Anus
- ☐ Liver
- ☐ Pancreas
- ☐ Gallbladder
- ☐ Appendix

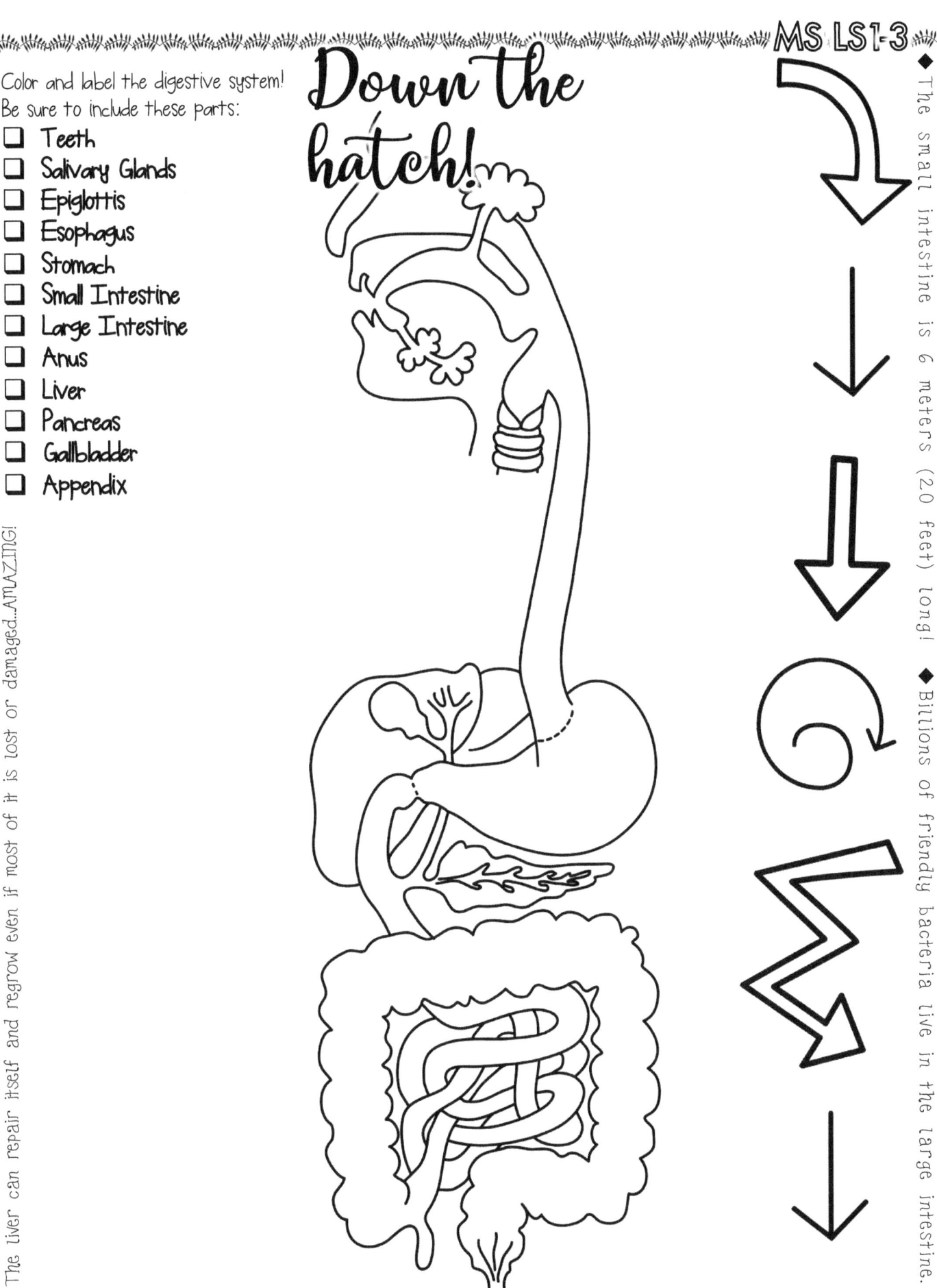

◆ The small intestine is 6 meters (20 feet) long!

◆ Billions of friendly bacteria live in the large intestine.

The liver can repair itself and regrow even if most of it is lost or damaged...AMAZING!

◆ The pancreas produces insulin which regulates sugar levels in the blood.

© 2021 Captivate Science

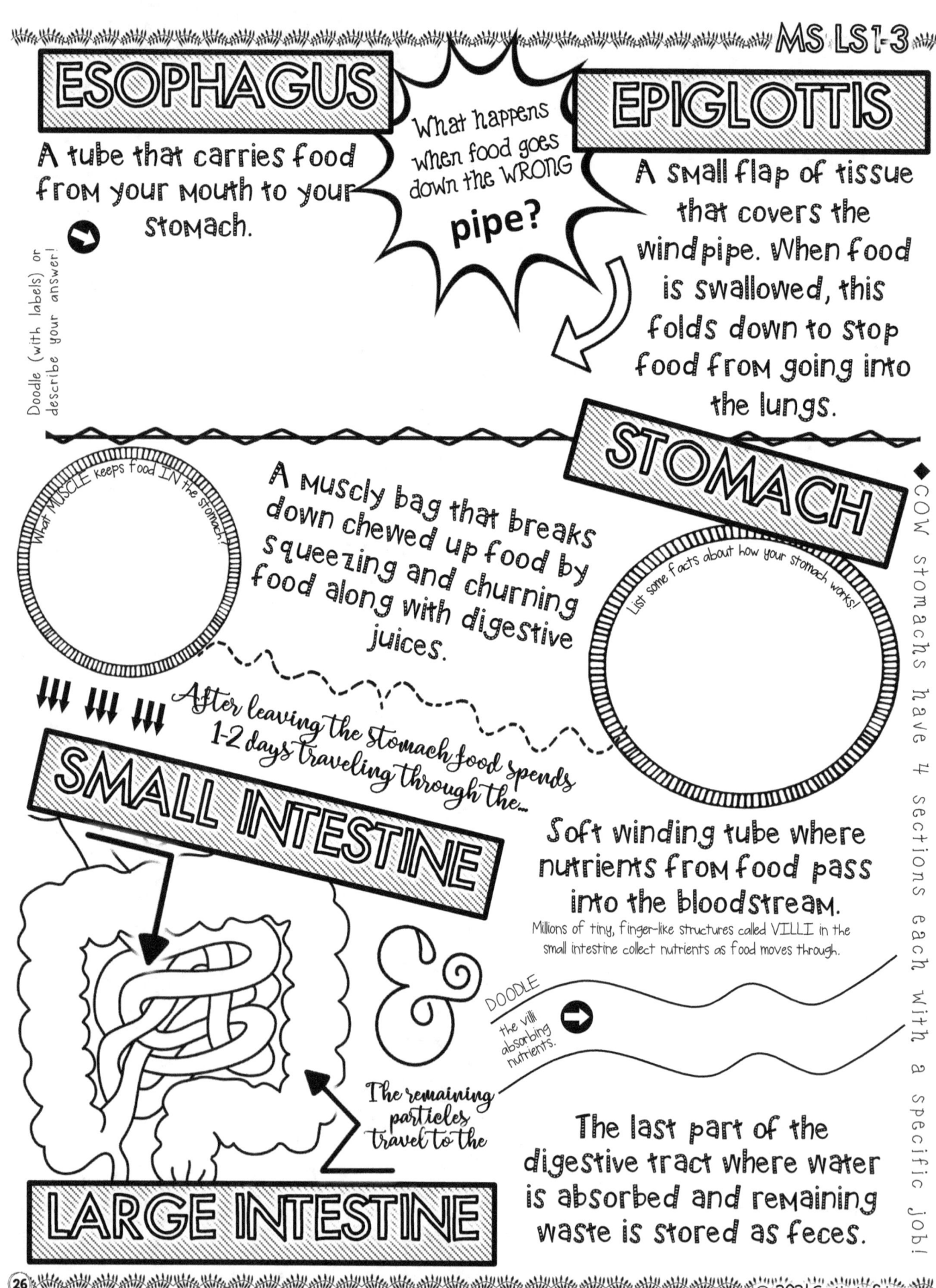

# Digestion Team!

**MS LS1-3**

These essential organs help the intestines digest food.

## LIVER

The largest and heaviest organ!

## GALLBLADDER

Bile created in the liver is stored in a small bag called the gallbladder. Bile is a green liquid that is passed to the small intestine to help digest fat.

Once nutrients have been absorbed in the small intestine, they are passed to the liver.

The liver has hundreds of different processes. List 3 things that it does to support digestion.

① 

② 

③ 

## PANCREAS

Long, flat gland behind the stomach that produces enzymes and hormones. It helps break down carbohydrates, fats and proteins.

→ MEMORY TRIGGER ←

→ In the space above DOODLE or DESCRIBE how YOU will remember the role of the liver, pancreas or gallbladder. ←

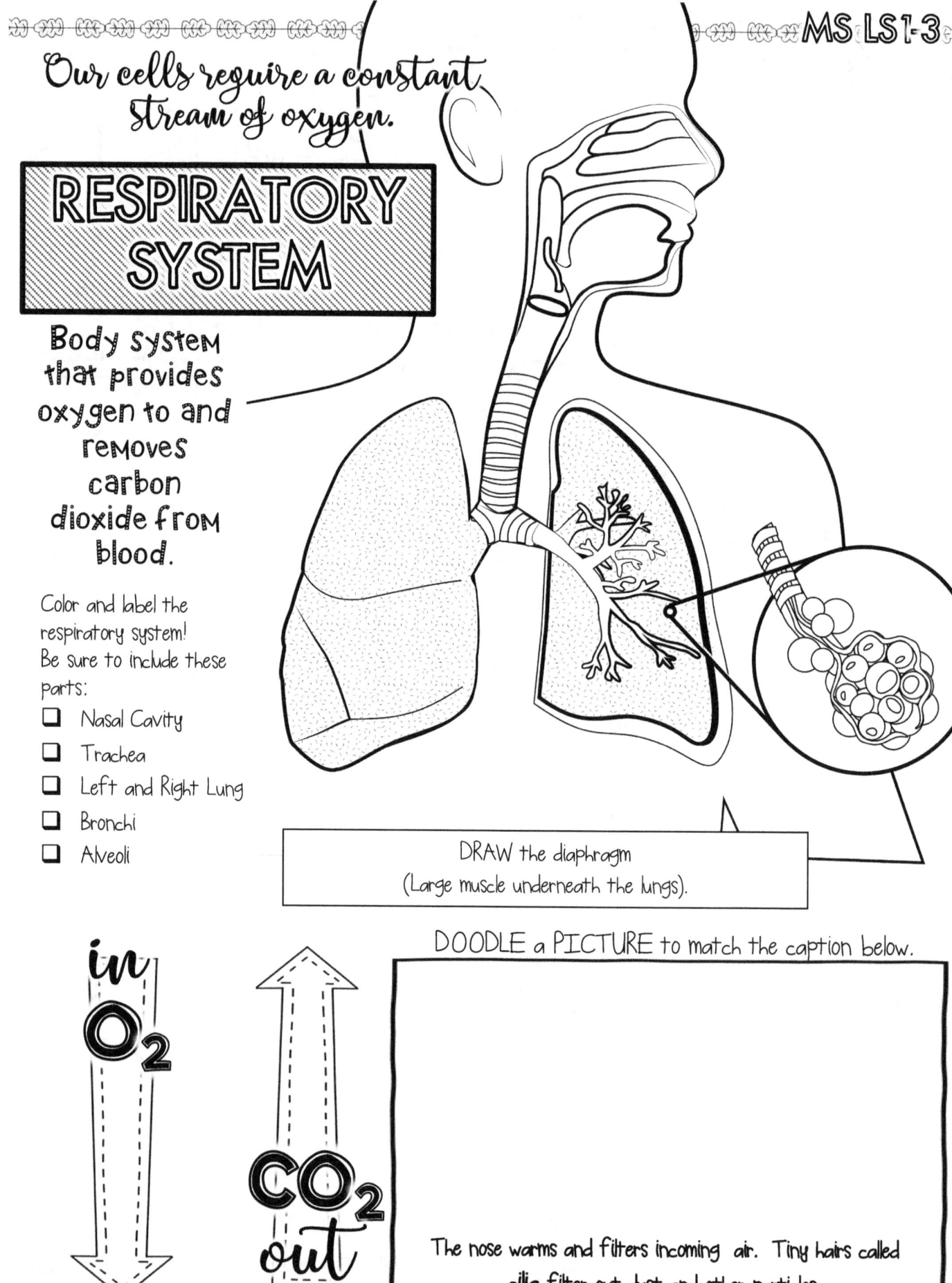

MS LS1-3

Our cells require a constant stream of oxygen.

## RESPIRATORY SYSTEM

Body system that provides oxygen to and removes carbon dioxide from blood.

Color and label the respiratory system! Be sure to include these parts:
- ☐ Nasal Cavity
- ☐ Trachea
- ☐ Left and Right Lung
- ☐ Bronchi
- ☐ Alveoli

DRAW the diaphragm (Large muscle underneath the lungs).

in $O_2$

$CO_2$ out

DOODLE a PICTURE to match the caption below.

The nose warms and filters incoming air. Tiny hairs called cilia filter out dust and other particles.

MS LS1-3

A.K.A "voice box"

## LARYNX

The larynx has vocal cords that vibrate as the air moves past them. This produces the sound of your voice.

Draw a FLOWCHART showing how air travels **through the respiratory system** when you breathe in.

## TRACHEA

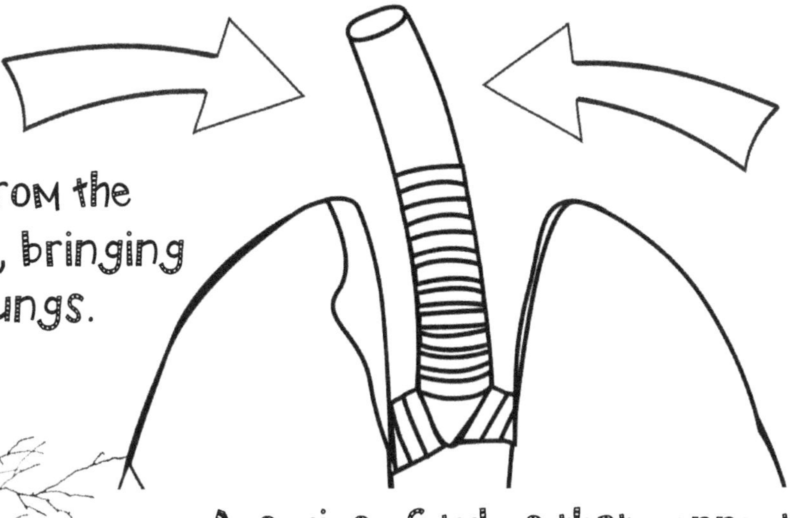

Tube than extends from the larynx to the bronchi, bringing air to and from the lungs.

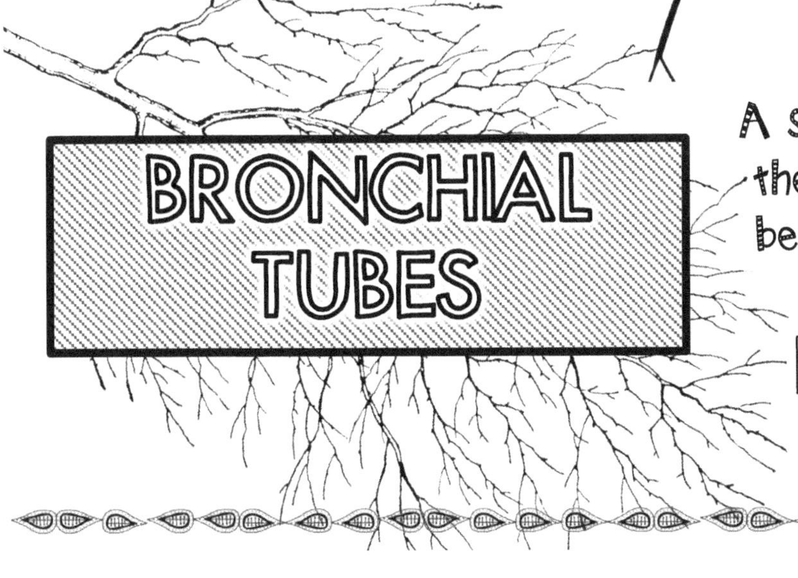

## BRONCHIAL TUBES

A series of tubes that connect the throat to the lungs. They become smaller as they travel deeper into the lungs.

Like BRANCHES on a TREE!

MS LS1-3

## LUNGS

A sponge-like organ in our chest that transports oxygen from the air you breathe into your bloodstream.

The lungs also remove carbon dioxide waste from the body when we breathe out!

**DOODL** a picture of the lungs that shows what you know!

### Did you know?

YOUR lungs have the surface area of a tennis court!

You inhale about 2,500 gallons of air each day!

Ultra-thin walls made of a single layer of cells allow for the <u>exchange of gases</u> between the air and the blood.

## ALVEOLI

Small sacs at the end of the bronchioles. The alveoli are wrapped in capillaries. Oxygen is passed from the alveoli to the capillaries.

bronchi
capillary
alveoli

Red blood cells
capillary
ultra-thin wall
ALVEOLUS
$O_2$  $CO_2$

Hemoglobin in our red blood cells has iron in it. The iron bonds with the oxygen in the air we breathe.

AIR

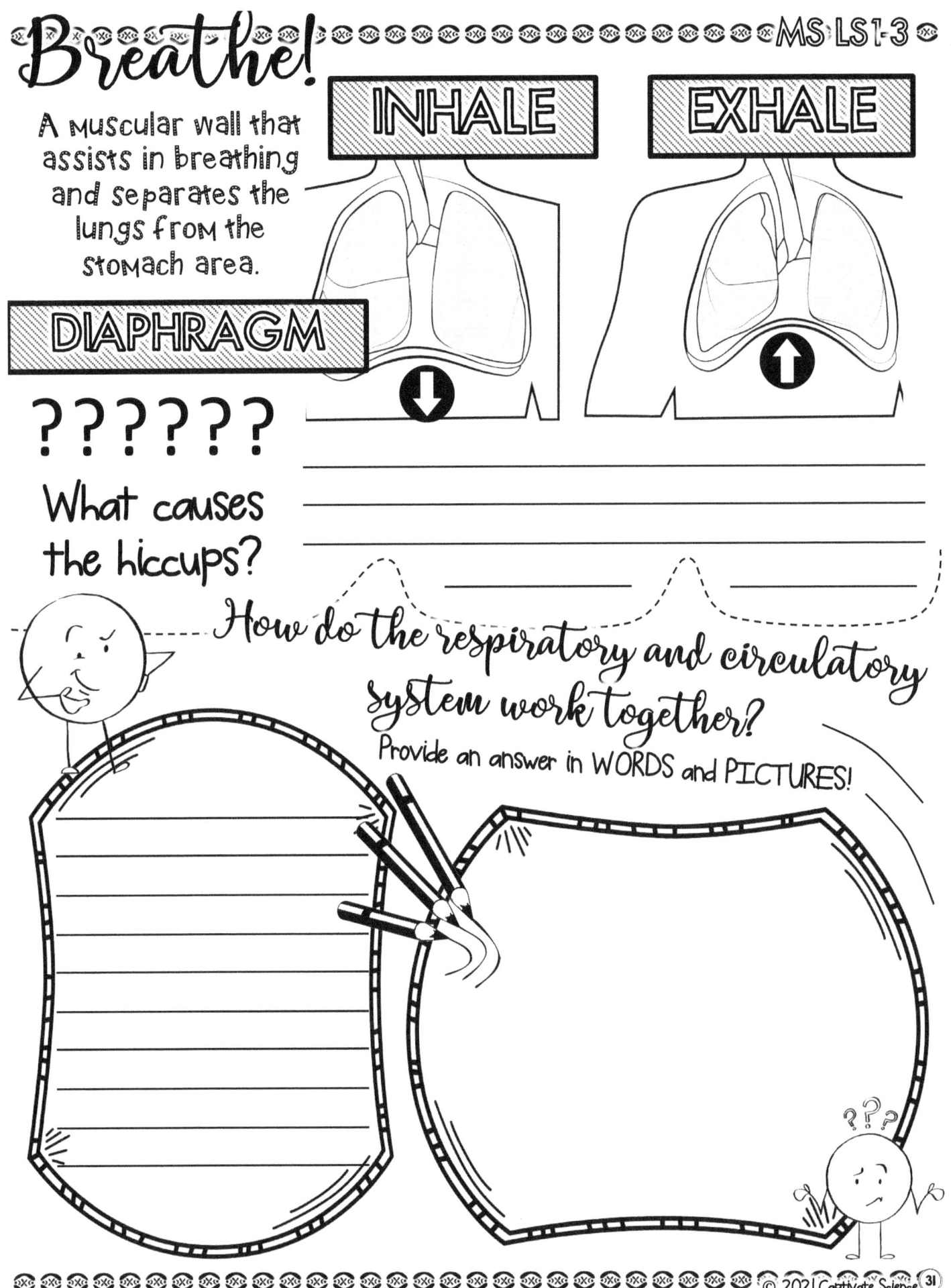

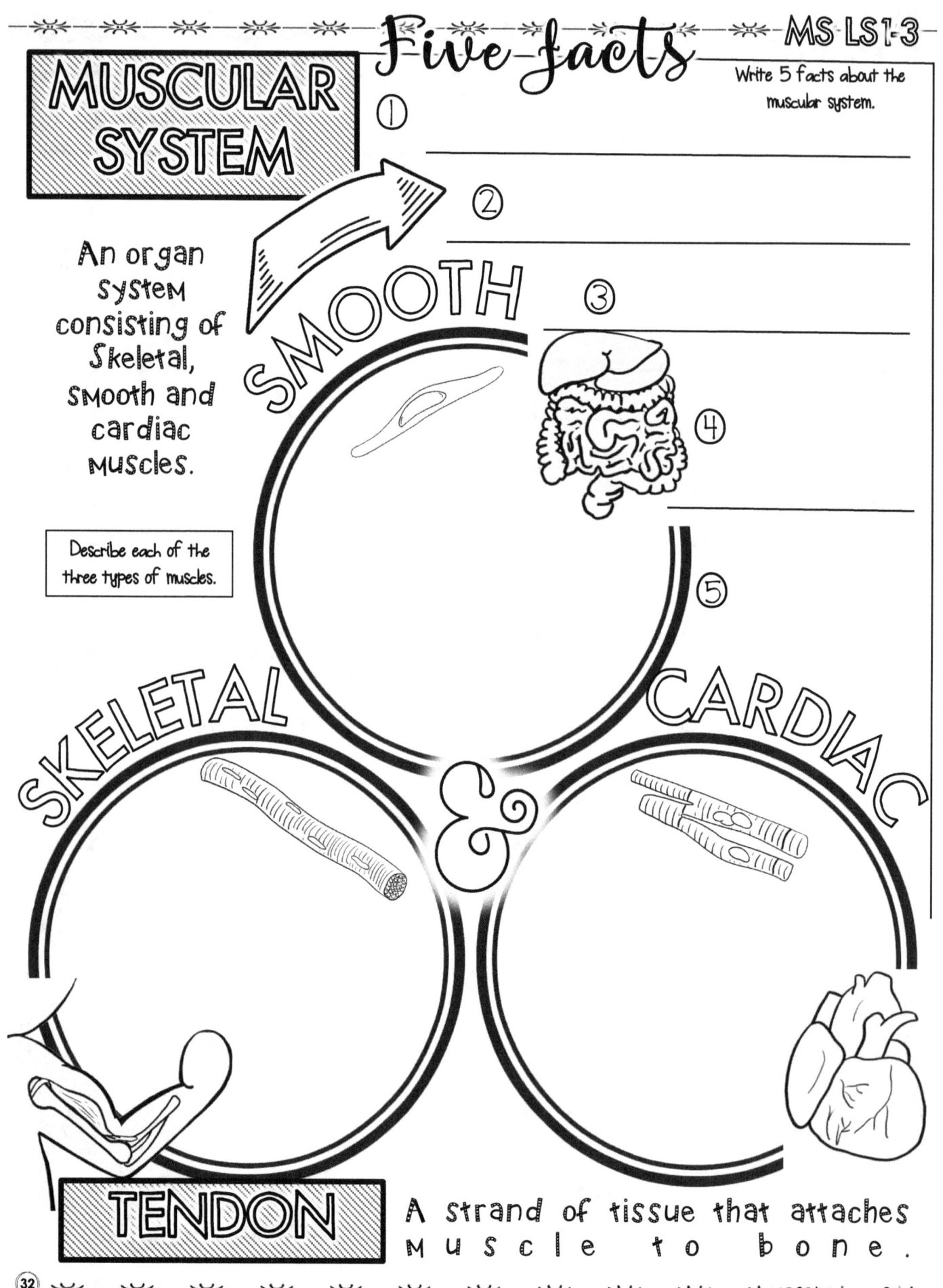

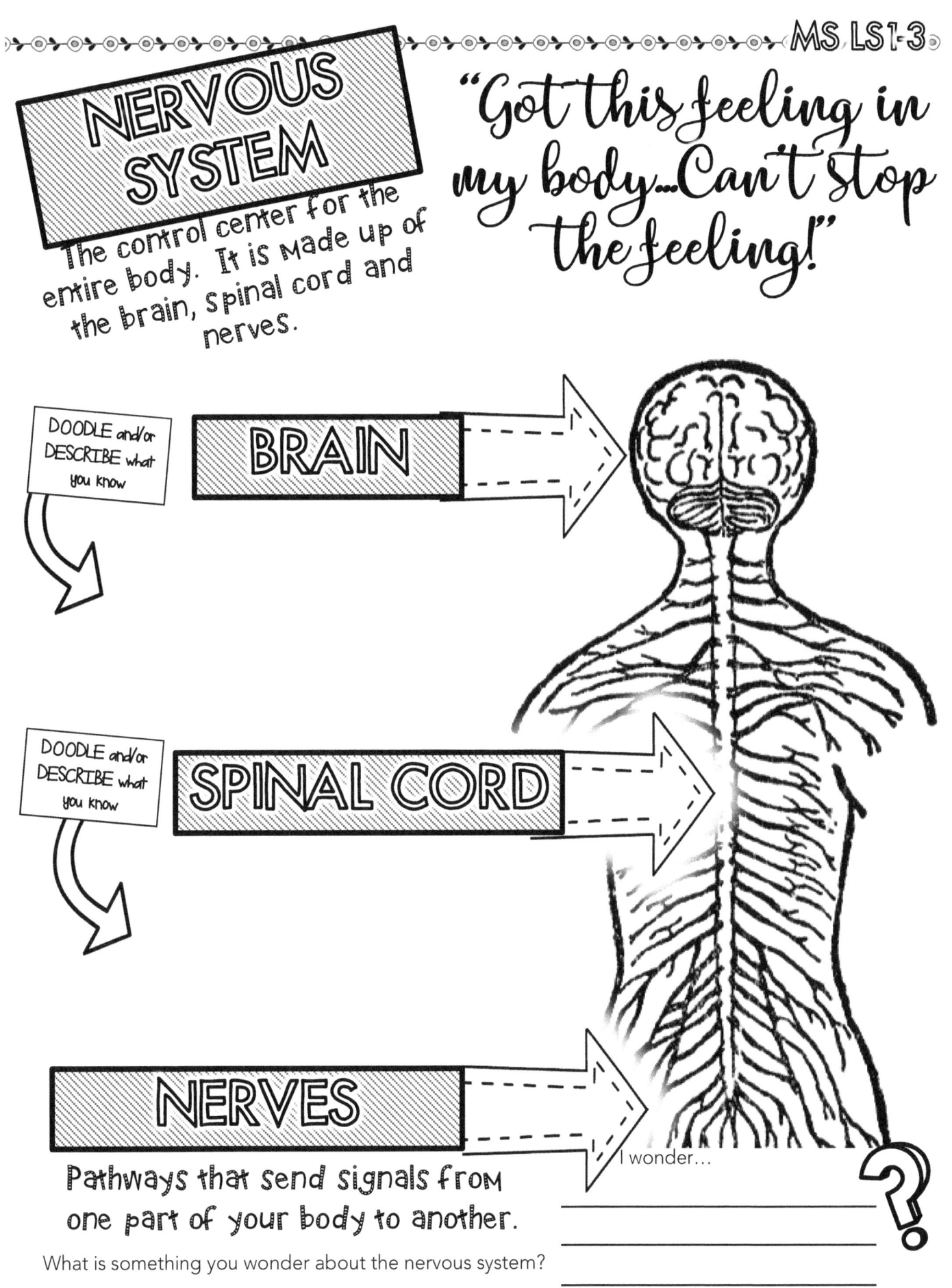

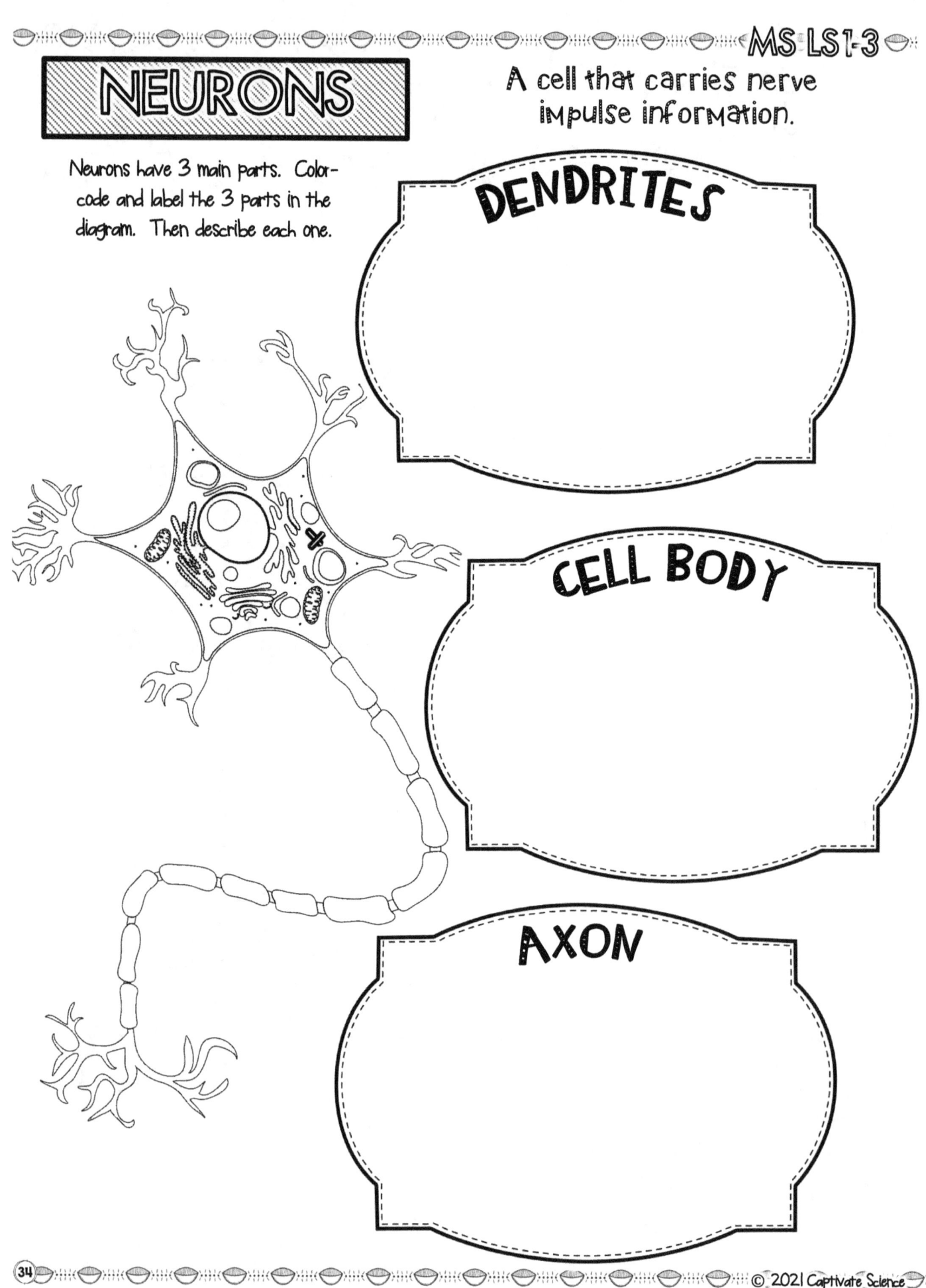

## SYNAPSE

The SPACE between two nerve cells where chemical messages are moved from the axon of one neuron to the dendrite of the next.

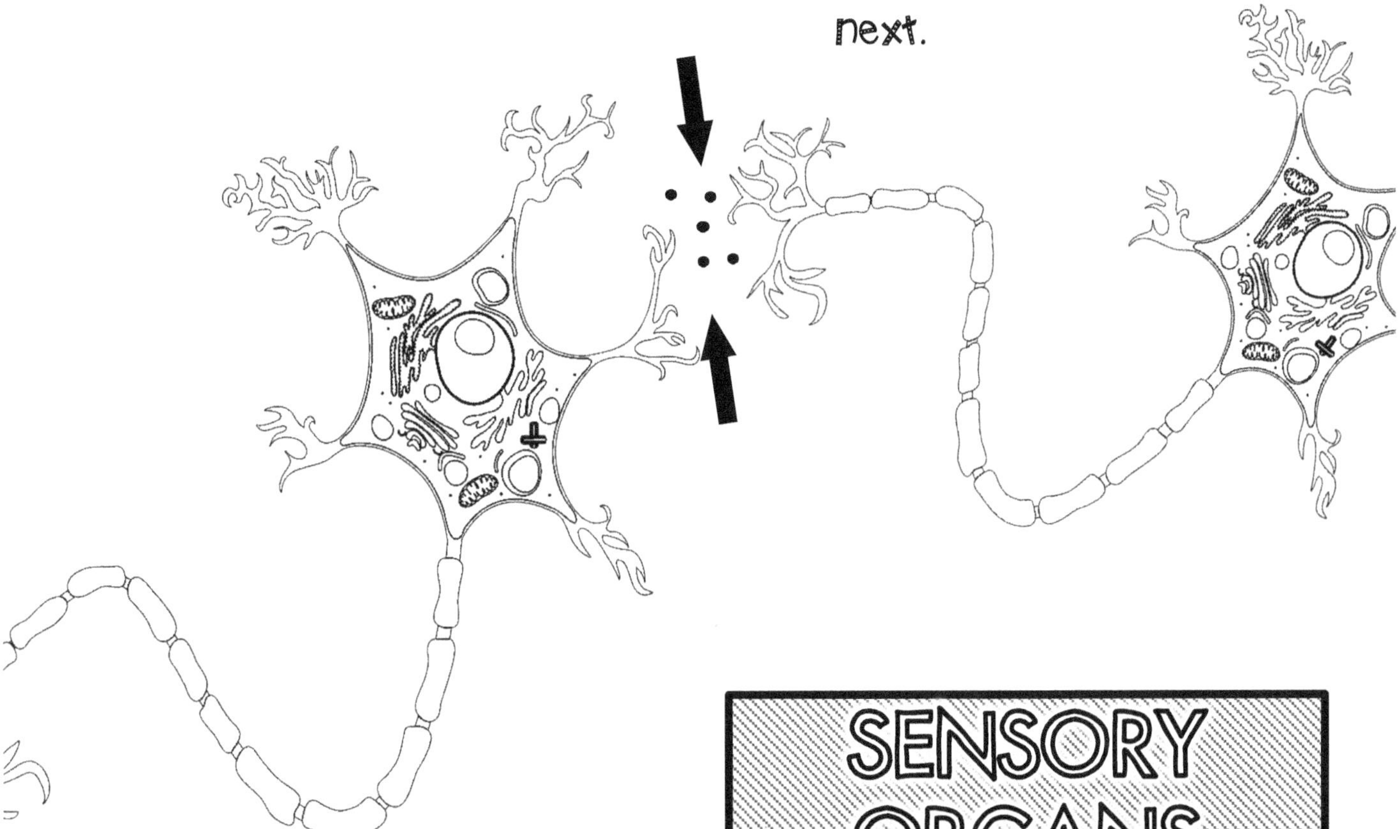

## SENSORY ORGANS

Specialized tissues within the skin, eyes, nose, ears and tongue receive stimuli and translate them into signals the nervous system can use. Nerves send the signals to the brain, which interprets them as sight, sound, smell, taste, and touch.

DOODLE a picture to match each caption.

| The eyes translate image signals into images for the brain to process. | Specialized skin tissue sends touch signals to the brain to process. | Draw your OWN example of a sensory organ sending information to your brain. |
|---|---|---|
| | | |

# MS LS1-3

## CENTRAL NERVOUS SYSTEM C.N.S
The brain and the spinal chord.

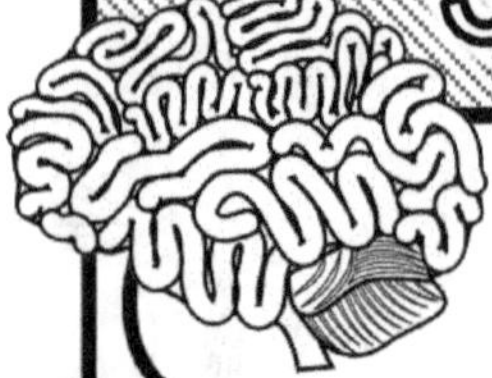

## PERIPHERAL NERVOUS SYSTEM
The sensory organs and the nerves that branch from the central nervous system (CNS) to the rest of the body.

In the diagram below, color code the CNS and the PNS.

CNS= LIGHT BLUE
PNS= RED

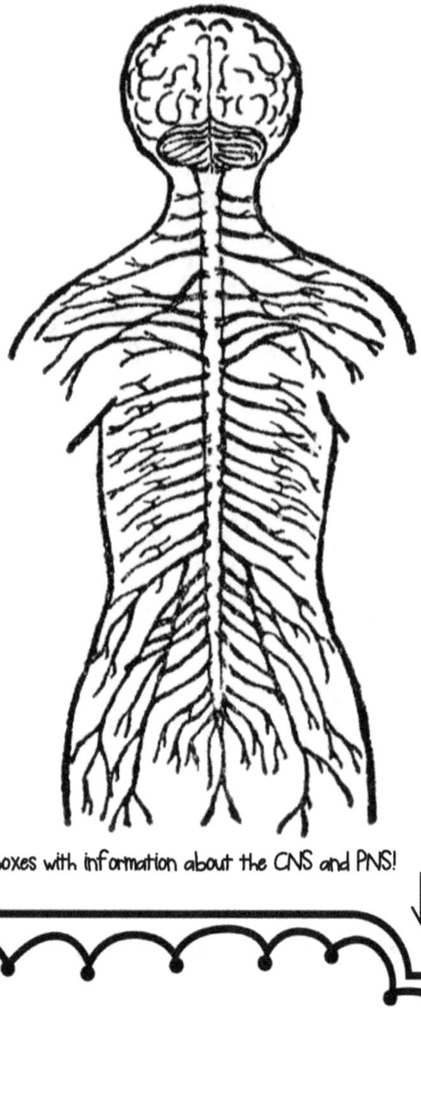

Fill these boxes with information about the CNS and PNS!

P.N.S

# How do organ systems work together?

MS-LS1-3

## CELLULAR RESPIRATION

The process of cells converting fuel into energy and nutrients.

GLUCOSE + OXYGEN → CARBON DIOXIDE + WATER + USABLE ENERGY

Name the body system responsible for each part of cellular respiration.

The _____ system breaks down food into nutrients (for example: glucose).
The _____ system brings oxygen to the lungs.
The _____ system transports the nutrients and oxygen to cells.
The _____ system brings carbon dioxide waste to the lungs.
The _____ system removes carbon dioxide from the body when we breathe out.

Most tasks require more than one body system working together. Which body systems are working together in each picture?

Lifting Weights

Eating Apples

_____

_____

Draw two more examples and label the body systems working together in your examples.

© 2021 Captivate Science

# SEXUAL REPRODUCTION

MS-LS1-4

Reproduction that requires fertilization and results in offspring that are genetically different from either parent.

Draw a set of labeled pictures to match the captions.

- Sunflowers are covered in a sweet pollen that attracts bees and other insects.
- When bees arrive, their feet get wet with pollen while drinking the sweet nectar.
- Pollen is transferred from the male (Anther) to the female part (Stigma).
- Seeds form in the center of the flower. These seeds fall out and regrow new plants in shallow soil.

What are some advantages and disadvantages of sexual reproduction? _____
_____
_____

# ASEXUAL REPRODUCTION

Reproduction that requires only ONE parent. The parent clones itself and offspring are genetically identical to the parent.

Color the picture and write a description of how strawberries reproduce.

Doodle and Label THREE other organisms that reproduce asexually.

MS LS1-4

## SEEDS

When an ovule is fertilized it divides and each fertilized cell becomes a seed.
An ovary of a plant can contain many seeds. A melon, for example, is an ovary with many seeds at its core.

As seeds mature, the ovary's smell, taste and appearance can attract animals.

YUM!

How do animals help plants reproduce?

Give an example of how animals help seed-producing plants reproduce.

## ANIMAL REPRODUCTIVE BEHAVIOR

ANY event or action that improves the ability of an organism to generate at least one replacement of itself.

## MATING

The pairing up of adult male and female animals to produce young. Animals that are successful at finding a healthy mate, are likely to have healthy offspring.

# Nature verses Nurture?

A.K.A "Genetics vs. the Environment"

MS-LS1-5

**GENETICS**

The branch of biology concerned with the study of heredity and variation of traits in organisms.

DOODLE a picture that helps you remember the meaning of the words on this page.

**HEREDITY**

The transmission of genetic characteristics from parent to offspring.

**GENE**

A unit of DNA that controls the development of one or more traits. (Genes are found in chromosomes).

## Nature's ingredient

**GENETIC FACTORS**

The parts of an organism that come from its parents.

Think about it in terms of cookies. Genetic factors are like ingredients. The ingredients make up the NATURE of your cookies.

What ingredients are in chocolate chip cookies?

The ingredients are like the genetic factors that influence living things.

What OTHER factors will influence how your homemade cookies taste?

These other factors are called environmental factors.

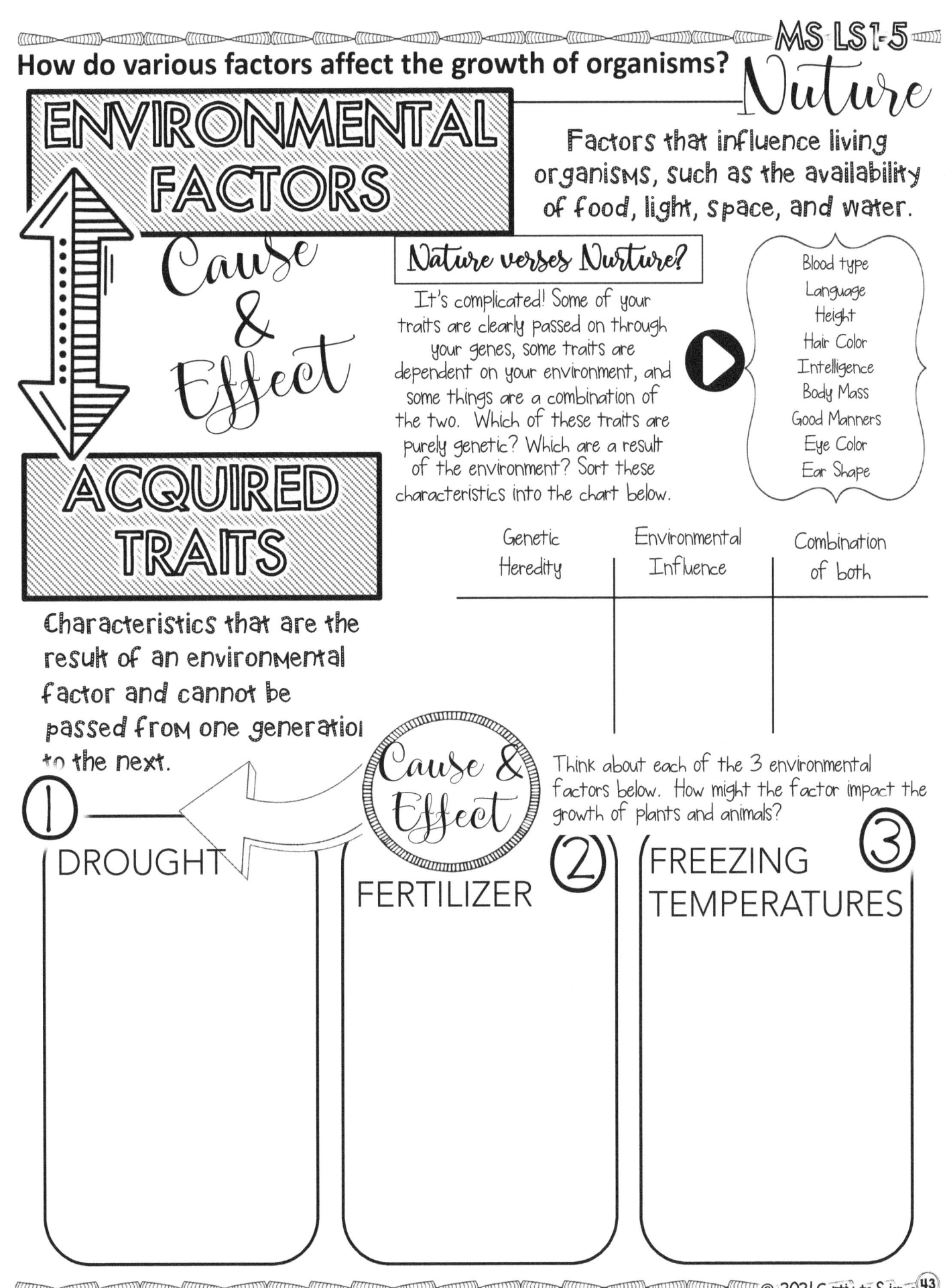

## How do environmental factors impact plant growth?

MS LS1-5

**OBSERVATION:**
Leaves near the bottom of an Oak tree have a larger surface area compared to leaves that are located near the top of the tree.

**TASK:** Using your background knowledge and research about photosynthesis and transpiration, write a claim supported by evidence and reasoning.

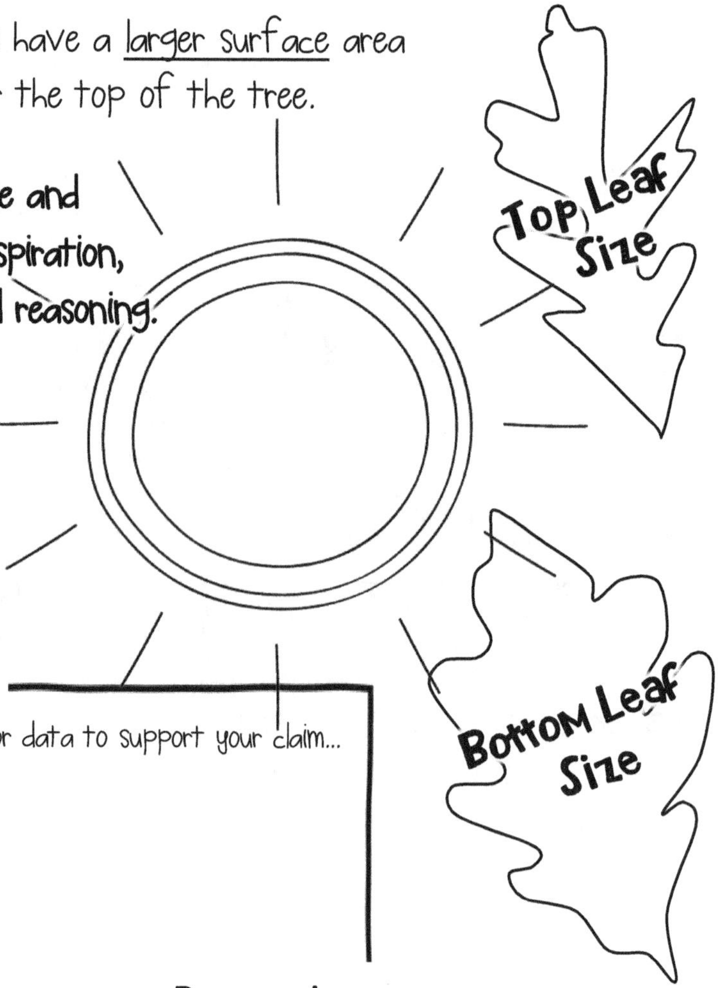

*Top Leaf Size*

*Bottom Leaf Size*

### Claim
The leaves are different sizes because...

### Evidence
List some facts or data to support your claim...

### Reasoning
The evidence supports my claim because...

List at least ONE source of information for your evidence.

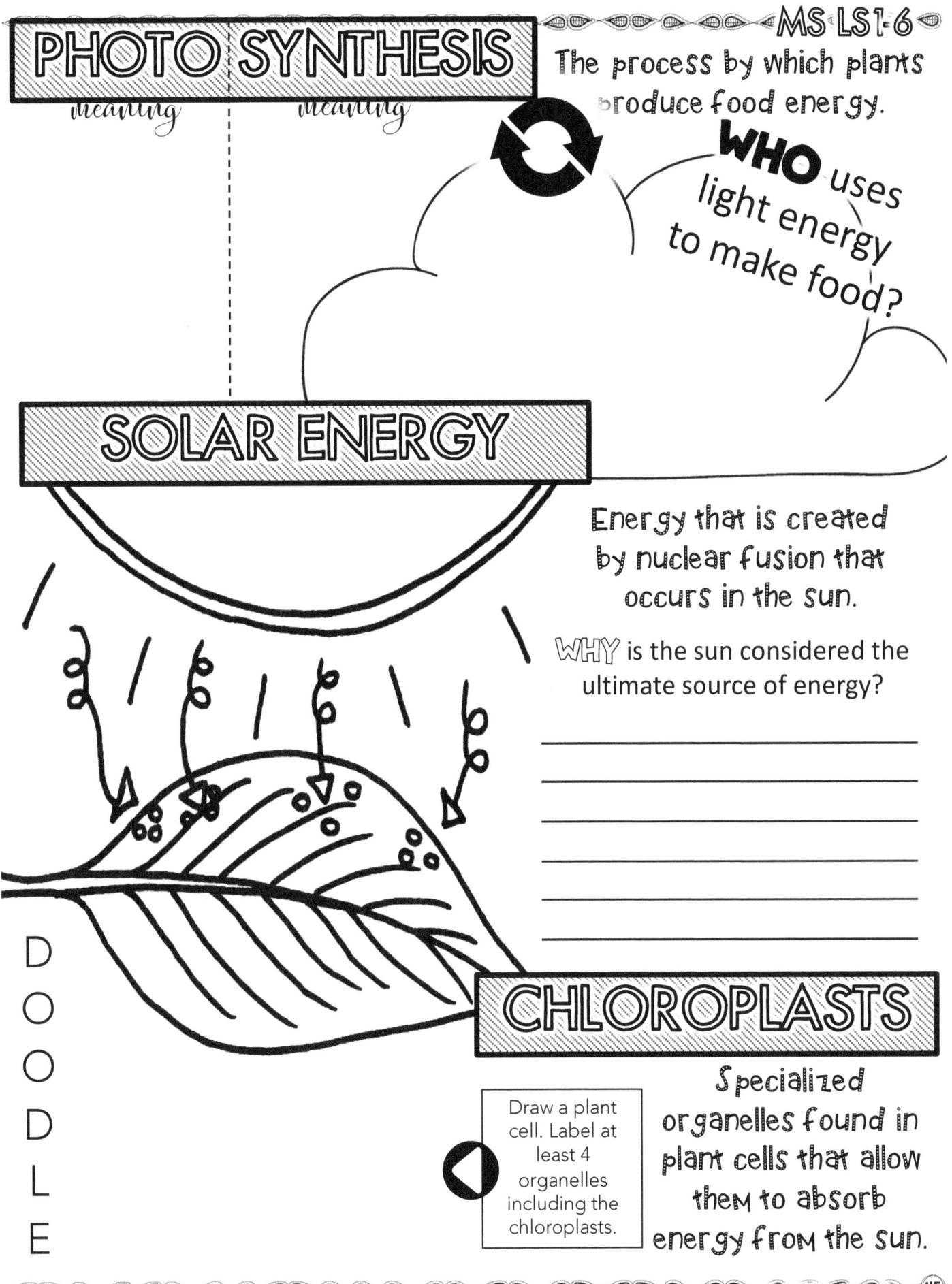

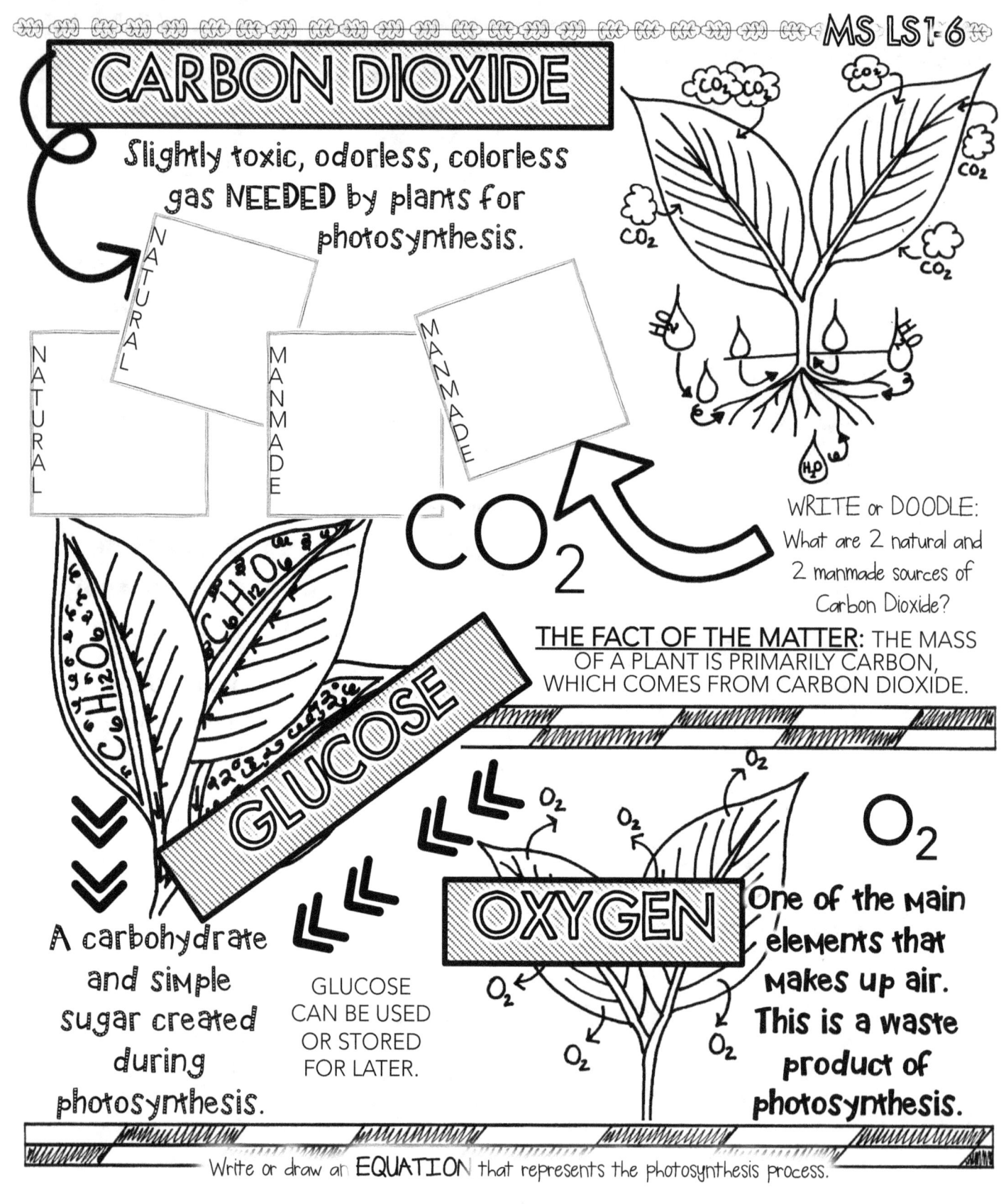

# CARBON DIOXIDE

Slightly toxic, odorless, colorless gas NEEDED by plants for photosynthesis.

NATURAL  NATURAL  MANMADE  MANMADE

$CO_2$

WRITE or DOODLE: What are 2 natural and 2 manmade sources of Carbon Dioxide?

**THE FACT OF THE MATTER:** THE MASS OF A PLANT IS PRIMARILY CARBON, WHICH COMES FROM CARBON DIOXIDE.

## GLUCOSE

A carbohydrate and simple sugar created during photosynthesis.

GLUCOSE CAN BE USED OR STORED FOR LATER.

## OXYGEN

$O_2$

One of the main elements that makes up air. This is a waste product of photosynthesis.

Write or draw an EQUATION that represents the photosynthesis process.

___ + ___ = ___ + ___

MS LS1-7

What 3 ELEMENTS are part of the photosynthesis and cellular respiration process?

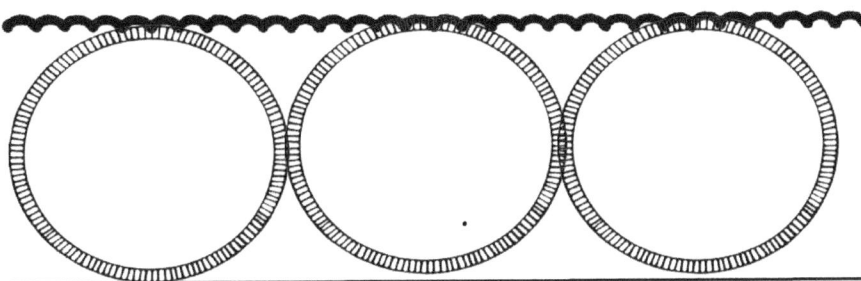

When these elements are arranged together what molecules do they make?

| SUBSTANCE | FORMULA |
|---|---|
|  |  |
|  |  |
|  |  |

PHOTOSYNTHESIS in chloroplasts

$CO_2$ + $H_2O$ WATER

FOOD $C_6H_{12}O_6$ + $O_2$

CELLULAR RESPIRATION in mitochondria

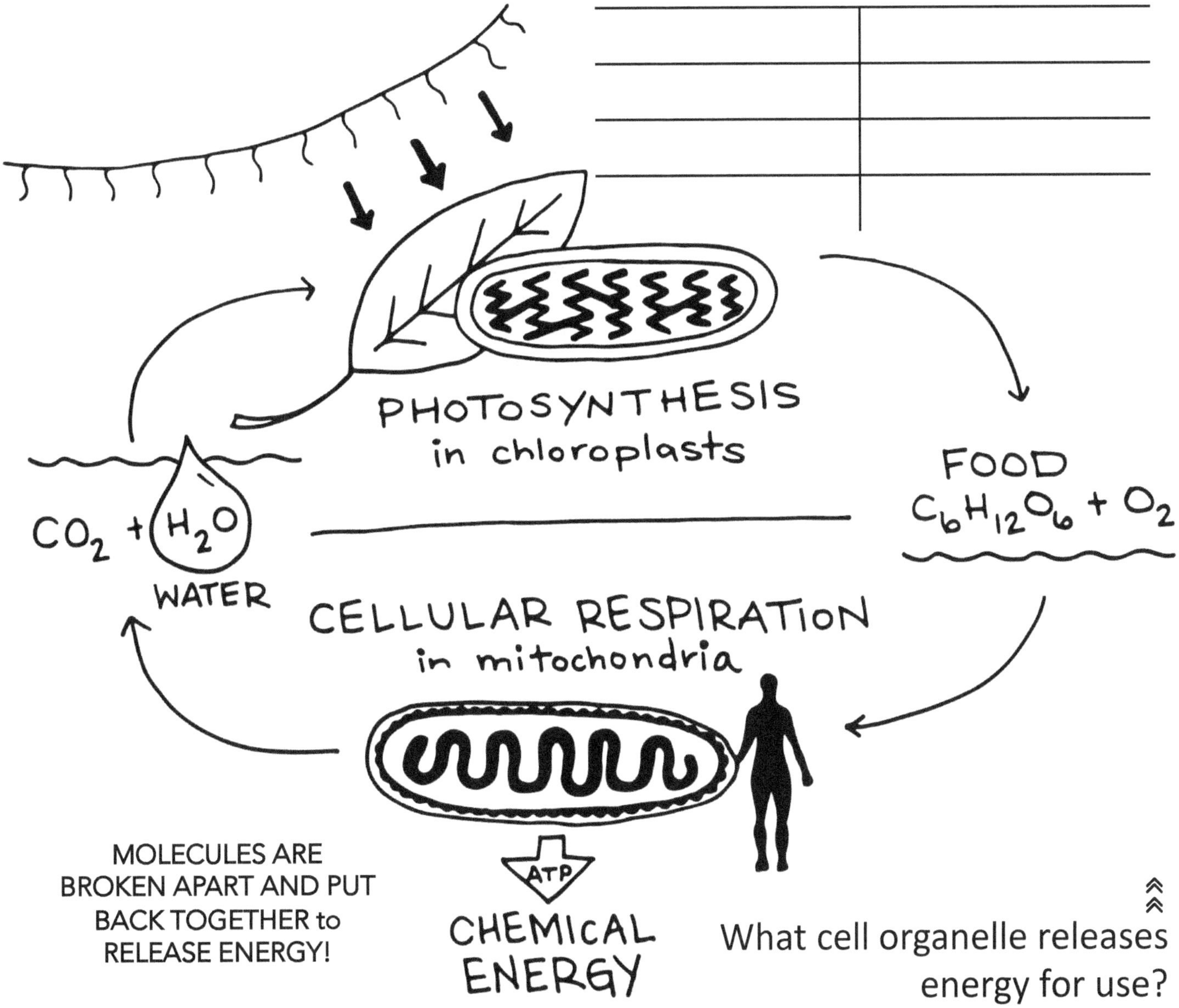

MOLECULES ARE BROKEN APART AND PUT BACK TOGETHER to RELEASE ENERGY!

ATP CHEMICAL ENERGY

What cell organelle releases energy for use?

**CELLULAR RESPIRATION**

The process by which organisms take oxygen and food molecules and convert them to energy and carbon dioxide.

MS LS1-7

# CHEMICAL REACTION

The process by which the atoms in a substance are reorganized to form a new substance.

FOR EXAMPLE: The chemical reaction for cellular respiration is:

GLUCOSE + OXYGEN ⇒ CARBON DIOXIDE + WATER + ATP

$C_6H_{12}O_6$    $O_2$          $CO_2$    $H_2O$    Energy

## OXYGEN

The gas that allows an organism's cells to make energy.

Just ∞ Breathe

## ATP   $

Short for "Adenosine triphosphate" which is a molecule that carries energy in cells. It is like energy currency on a cellular level.

DOODLE or DESCRIBE how a runner's leg muscles get energy for a race.

### ①②③ Sequence the steps below that explain how molecules are broken apart and put together to give organisms energy.
(Two of the steps have already been numbered for you!)

___ ATP is the energy that powers all the processes done by cells.

1   Plants and animals are interconnected!

___ Plants use sunlight to reorganize Carbon (C), Hydrogen (H) and Oxygen (O) to make glucose.

___ The mitochondria in our cells are powered by the oxygen we breathe.

___ We eat glucose as food and breath in oxygen- both produced by plants.

___ Mitochondria reorganize the molecules in oxygen and glucose to create ATP.

___ We breathe out Carbon Dioxide as a result of the work of our mitochondria.

8   Plants breathe in the Carbon Dioxide that we breathe out and use it to reorganize more molecules into glucose.

# How are MEMORIES formed?

MS LS1-8

## STIMULUS
An event or experience that causes an organism to react.

All memories begin with perception of an event or experience. We start by receiving information about the experience. Think about the first time you met an important person in your life (best friend, little brother...etc.) What did your senses gather that would help you remember the person?

← Make a list

## RECEPTOR
Any structure specialized to detect a stimulus. Examples of sense receptors:

### Electromagnetic Receptor
Detects _____

### Thermal Receptor
Detects _____

### Mechanical Receptor
Detects _____

### Chemical Receptor
Detects _____

© 2021 Captivate Science

# How are MEMORIES formed?

MS LS1-8

Once sense receptors gather information, we store that in our brains for short-term or long-term use. Most of what we experience day to day is used short term. For example, when you look up a number to order pizza you might recall it long enough to dial it but soon after you forget the digits. However, when we repeatedly use the same information our brain will store it for long term access. If you order pizza 3 days a week from the same place, you will likely remember the phone number!

## SHORT TERM MEMORY

Information for limited time and usage.

How can we move memories from short to long term?

Share a long term memory you have that you know you will not forget!

## LONG TERM MEMORY

Information that has been stored, organized and repeated for long term use.

### REFLEX
Simple response to a specific stimuli.

### INSTINCT

Patterns of behavior that exist in most members of a species.

Read each statement and determine whether it is a reflex or an instinctual behavior. Label reflex R and instinct I.

Pulling your hand away from heat  _____
Birds building a nest  _____
Babies knowing how to drink milk  _____
Pupils dilate to protect from light  _____

Think of another example and label it as a reflex or instinct.

MS LS1-8

## 3 FUN FACTS ABOUT YOUR BRAIN

Do some research about how your brain interprets information and stores it in your memory. Record 3 things you find.

① 

② 

③ 

How is your brain like a computer?

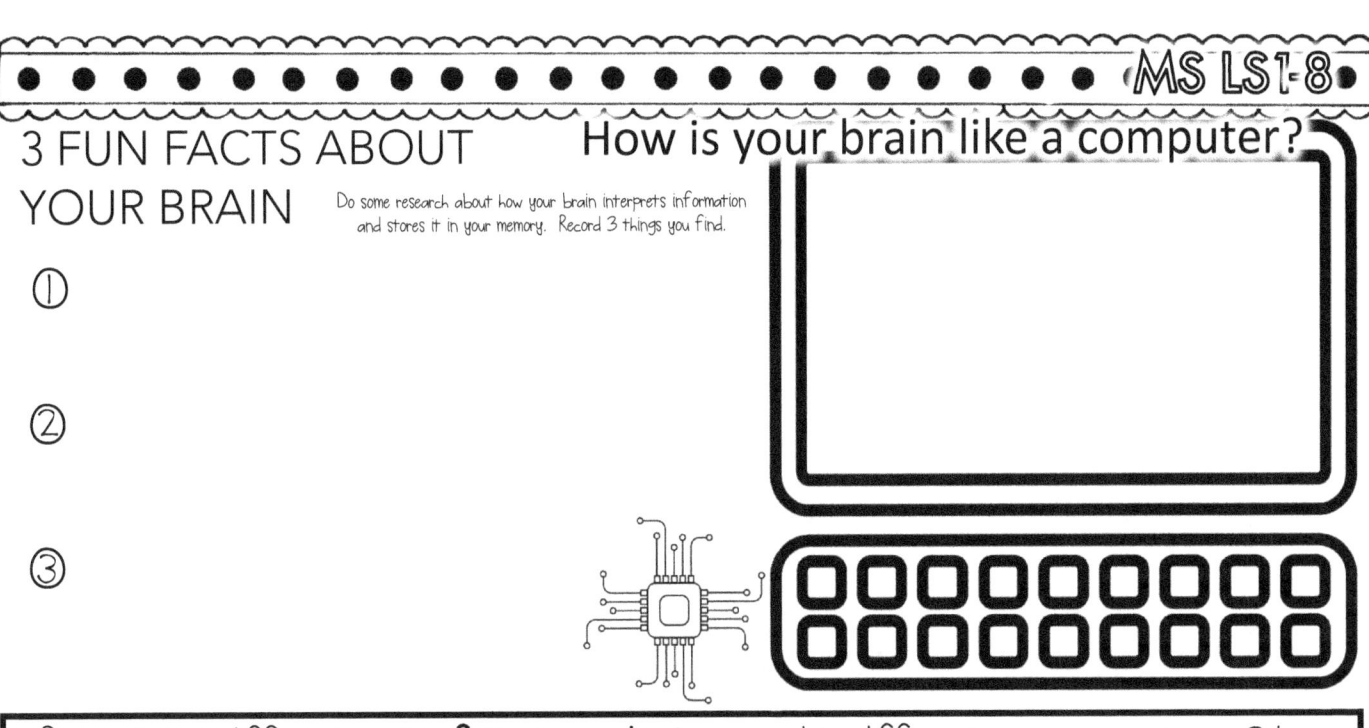

Research how different **parts of the human brain** respond to different sensory receptors. Color-code and label the brain below to show what you find out about different areas of the brain.

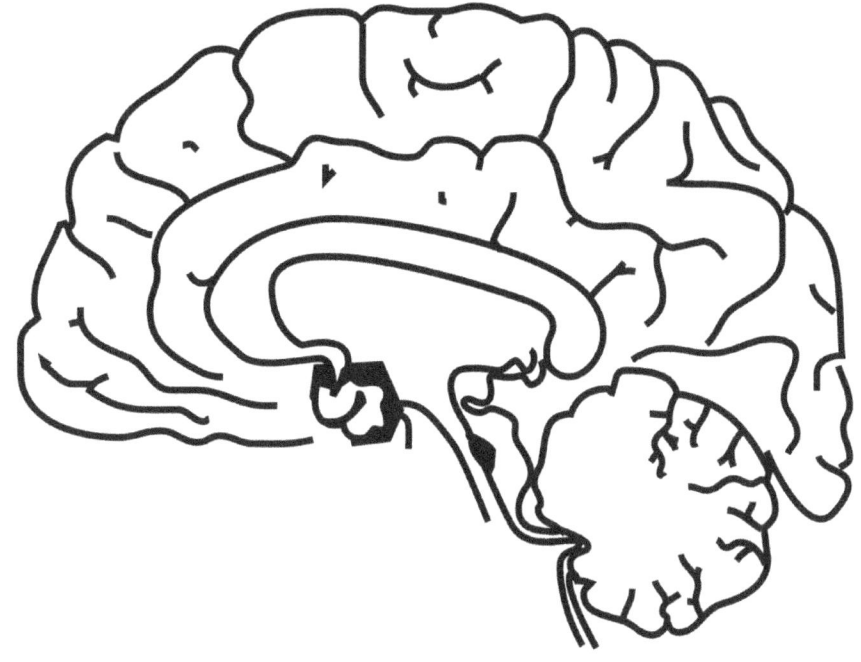

# CHAPTER 2

## Interactions, Energy, and Dynamics: Relationships in Ecosystems

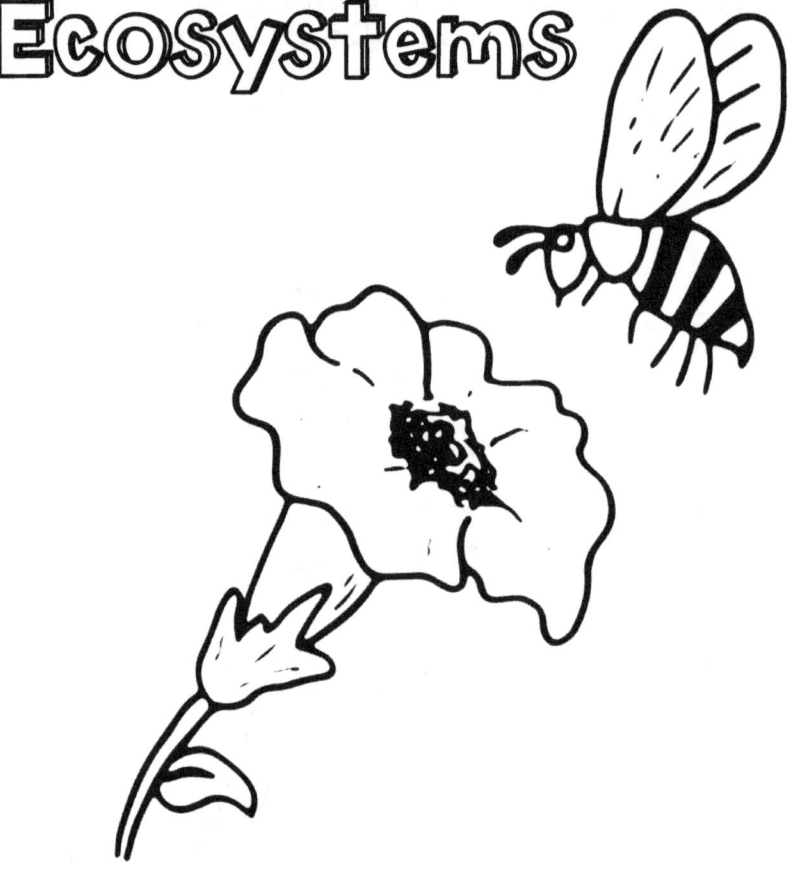

*How does a system of living and non-living things operate to meet the needs of the organisms in an ecosystem?*

# ORGANISM

Any living thing including a plant, animal, bacterium, virus or fungus.

MS-LS2-1

What makes something living vs. non-living? Explain your ideas here:

# POPULATION vs. COMMUNITY

A group of organisms of the **same** species living in the **same** place at the **same** time.

Populations of different species (plants and animals) living and interacting in the same area.

# ECOSYSTEM

A community of plant and animal populations PLUS the abiotic (non-living) environmental factors. This includes air, water, soil, temperature...etc.

SUMMARIZE: WHAT IS THE DIFFERENCE BETWEEN AN ORGANISM, A POPULATION, A COMMUNITY AND AN ECOSYSTEM?

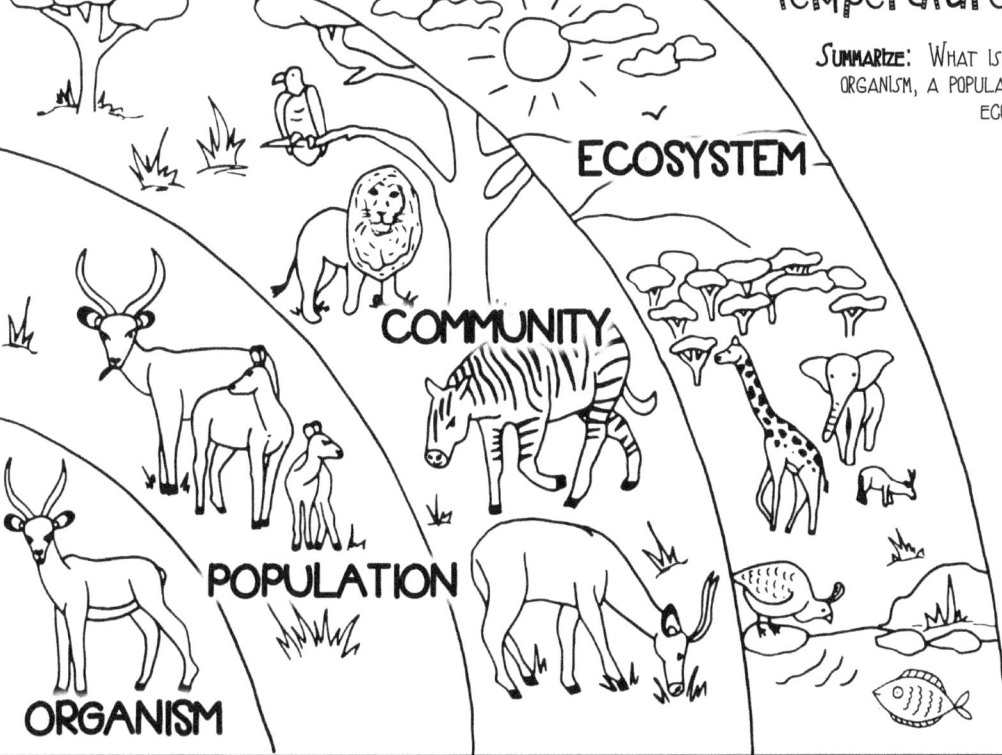

# INTERDEPENDENCE

MS-LS2-1

DOODLE a picture of how plants and animals are interdependent.

The way all organisms in an ecosystem depend upon each other. If the population of one organism increases or decreases, this can impact the rest of the ecosystem.

How does **resource availability** impact population size?

Use the graph below to answer questions Ⓐ – Ⓓ

Ⓐ Describe how the amount of snow changed over time.

Ⓑ Describe how the Cottontail rabbit population changed over time.

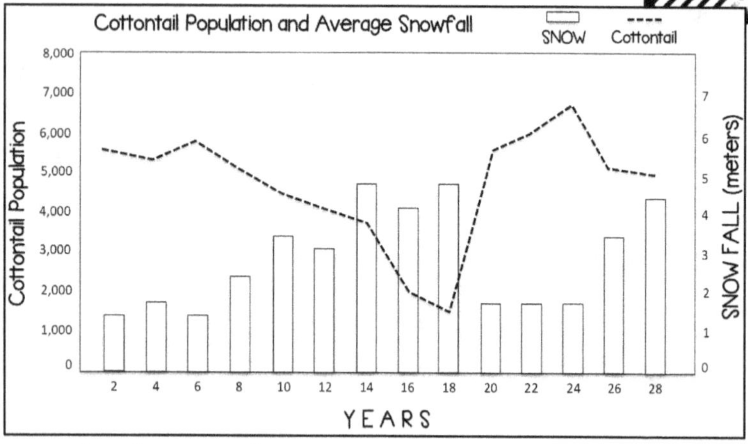

**populations increase**
① 
②

**populations decrease**
① 
②

Give **two** reasons why the population of a species might increase and two reasons why it might decrease.
① 
②

Ⓒ Does there appear to be a RELATIONSHIP between snow amount and rabbit population? If so, how do they relate?

Cottontail Population and Average Snowfall
— SNOW   ---- Cottontail

Ⓓ What do you predict would happen to the cottontail rabbit population if the snowfall amounts decreased over the next few years? Use evidence from the graph and scientific concepts to support your claim.

_____
_____
_____
_____
_____

54  ©2021 Captivate Science

# PREDATOR - PREY relationship

An interaction between two organisms of different species in which one of them (PREDATOR) feeds on the other organism (PREY).

*Think like a scientist.*
*evaluate* *explore* *analyze* *interpret*

MS-LS2-2

① ② ③ ④ ⑤ Describe the relationship between predator and prey populations of mice and hawks using the graph below.

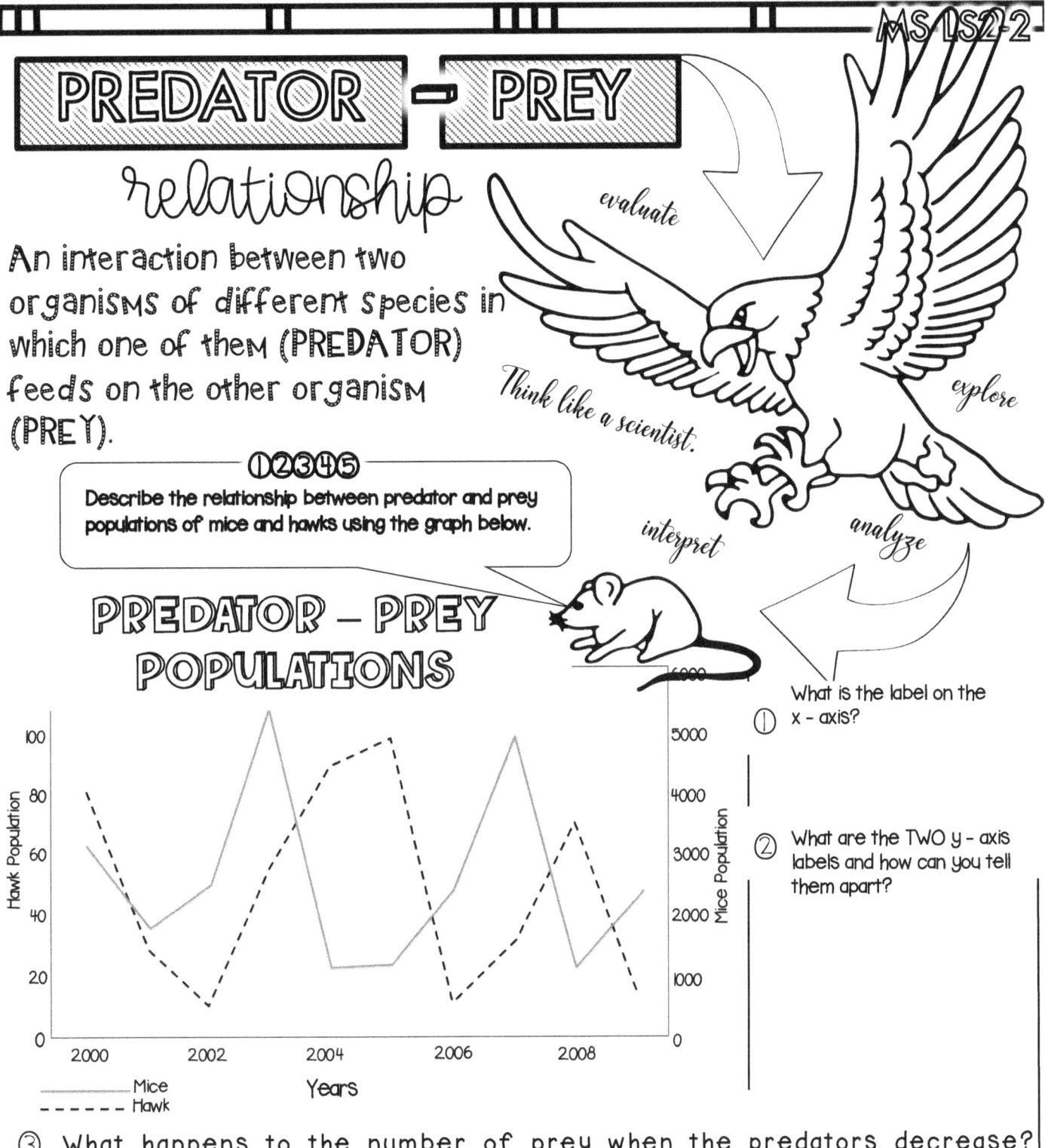

## PREDATOR - PREY POPULATIONS

(Graph showing Hawk Population and Mice Population vs. Years 2000–2008)
— Mice
- - - - Hawk

① What is the label on the x-axis?

② What are the TWO y-axis labels and how can you tell them apart?

③ What happens to the number of prey when the predators decrease?

④ Give a specific example from the graph of this relationship:

⑤ Explain (using a science concept) why this relationship happens:

MS·LS2-2

## MUTUALISM

A symbiotic relationship that is beneficial to BOTH organisms.

Explain how bees and flowers mutually benefit each other:

_____
_____
_____
_____
_____

## COMMENSALISM

A symbiotic relationship where one species benefits and the other is not impacted at all.

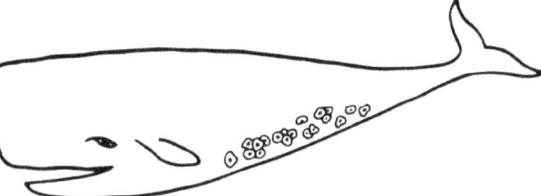

How do whales and barnacles demonstrate commensalism?

_____
_____
_____

## PARASITISM

Explain how a flea and a dog are an example of parasitism.

A symbiotic relationship where one species benefits and one species is HARMED.

_____
_____
_____
_____
_____

©2021 Captivate Science

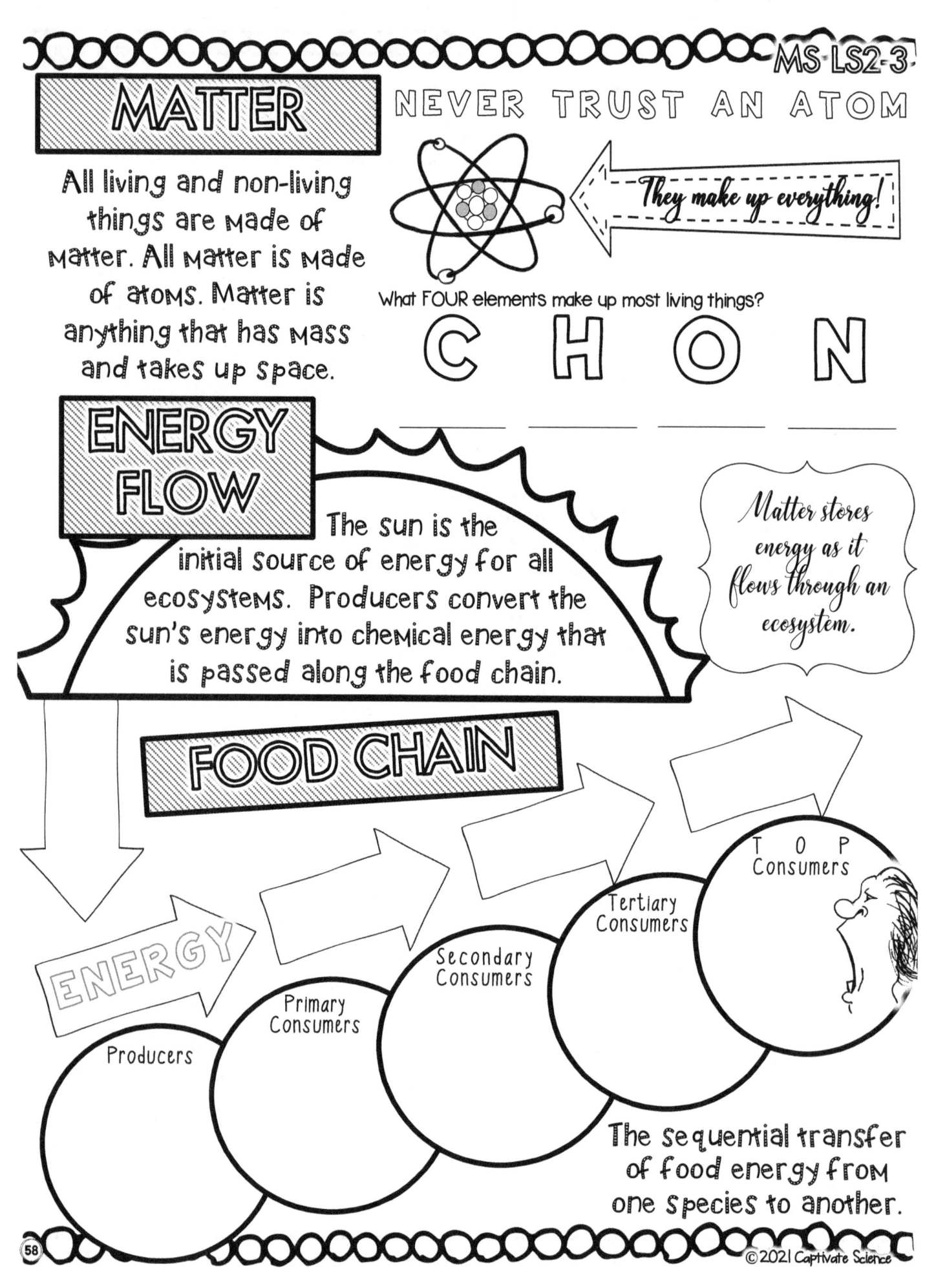

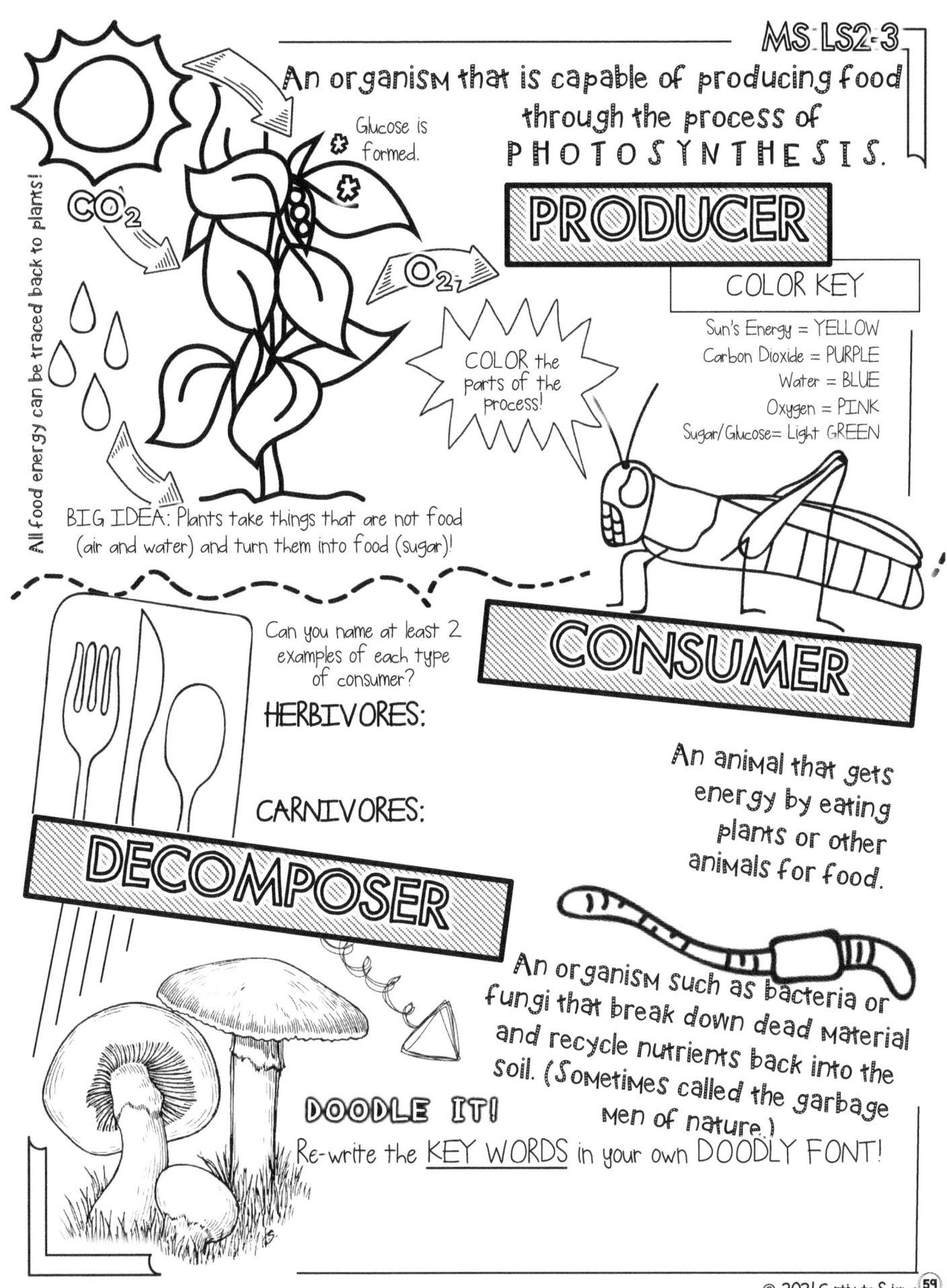

MS LS2-3

**NOTE:** Energy is not recycled, nor can it be created or destroyed.

**NOTE:** Energy is transformed from one form to another.

**NOTE:** Energy is released AS it flows through a food chain. (It is transferred to heat, kinetic and sound energy).

# ENERGY in an ECOSYSTEM

Food energy gets used by animals for life processes and then transfers to other forms of energy such as heat. This "used" energy is not passed on to the next animal.

10 Calories
100 Calories
1,000 Calories
10,000 Calories

Doodle some pond fish here.
Fill this section with doodles of snails and tadpoles.
Fill this section with pond plant doodles.

③ The energy is passed on to carnivore (meat eating) consumers when they eat another animal.

② The plant energy is passed to animals when herbivore (plant eating) consumers eat plants.

① Energy from the sun is captured by plants that make their own food energy.

# 10% ENERGY RULE

What is the other 90% used for?

ONLY about 10 per cent of the chemical energy is passed from one level of a food chain to the next.

**WHAT IF** You are stranded on an island with a large bag of grain and two egg-laying chickens. Mathematically speaking, what is your best eating strategy to have enough energy for survival? Use the 10% Energy RULE to support your claim.

THINGS THAT MAKE YOU GO HMMM...

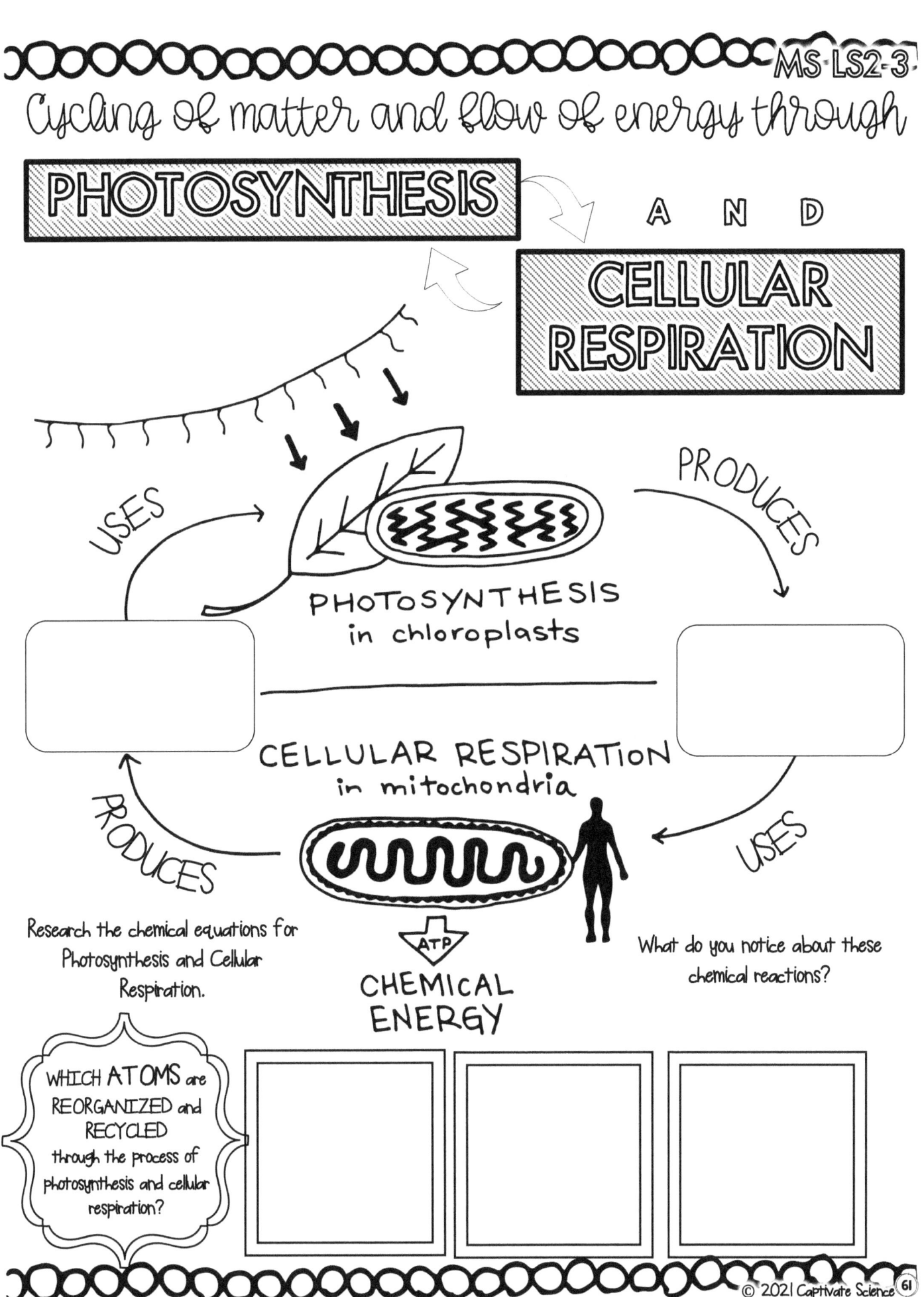

# FOOD WEB

MS LS2-3

A series of organisms connected by predator-prey and consumer-producer relationships. It includes many <u>overlapping food chains</u> to represent how energy moves through an ecosystem.

Examine the above food web.
Label it using the key.
(Note: Some organisms may have more than one label.)

**KEY**
P = producer    1 = primary consumer
2 = secondary consumer   3 = tertiary consumer
4 = quaternary consumer   H = herbivore
C = carnivore   O = omnivore   D = decomposer

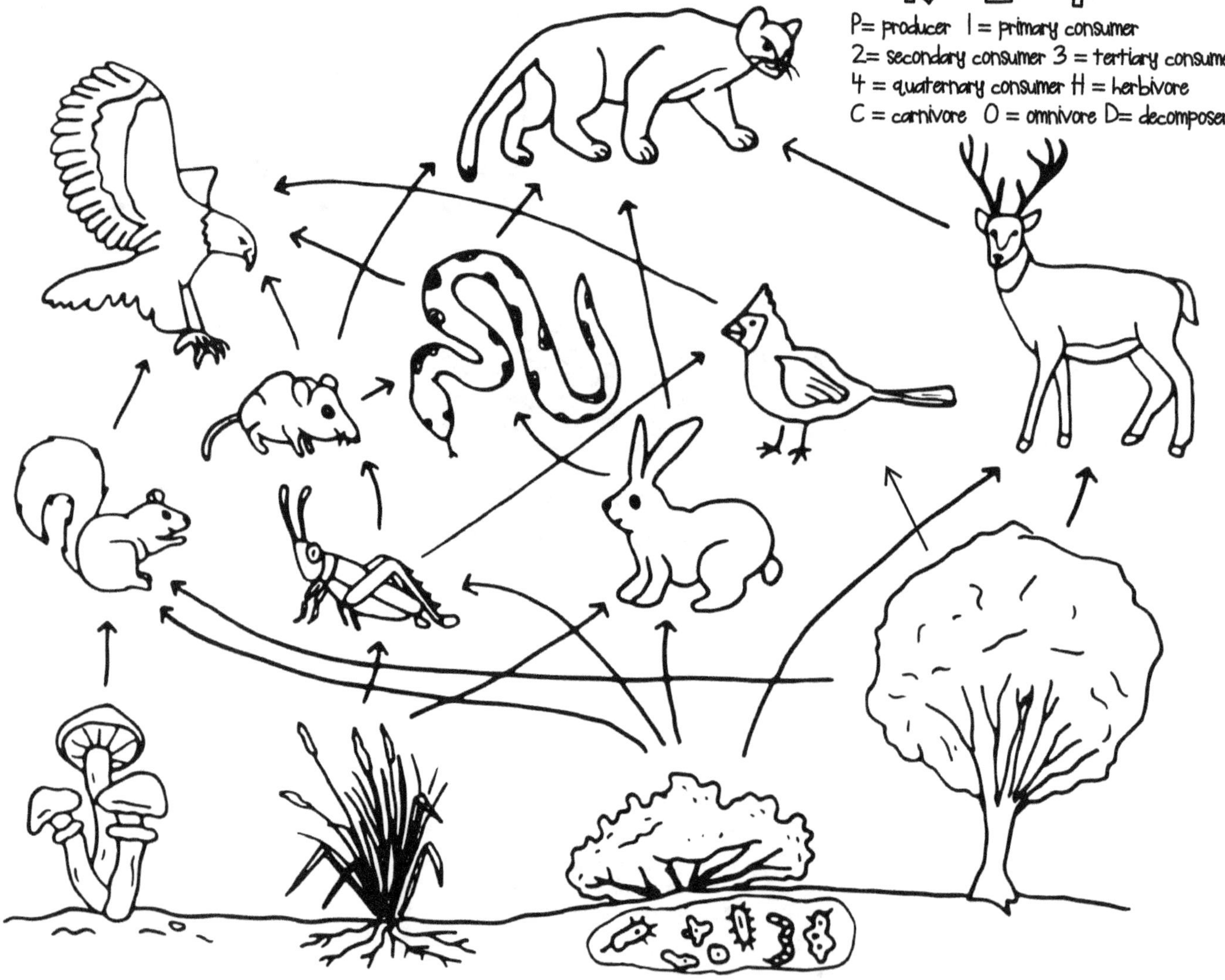

Can you identify some of the food chains in this food web?

## FOOD CHAIN #1    FOOD CHAIN #2    FOOD CHAIN #3

MS-LS2-4

## ABIOTIC FACTORS

Non-living chemical and physical parts of an environment that impact the function of an ecosystem.

### GIVE 3 REASONS
Why are abiotic factors important in an ecosystem?

1.
2.
3.

DOODLE as many abiotic factors as you can!

## ECOSYSTEM DISRUPTION

Any change that disturbs the natural balance of an ecosystem. All living and non-living parts of an ecosystem play a role in keeping an ecosystem healthy. If one aspect is disturbed, the rest of the ecosystem is impacted.

SORT these disruptions...

| Natural Disruptions | Human Disruptions |
|---|---|

Temperature — Logging
Invasive Species — Hunting
Air Pollution — Construction
Volcanic Eruption — Flooding
Water Diversion — Farming
Forest Fire — Disease

* Some might be classified as BOTH

**CAUSE AND EFFECT**

Pick one ecosystem disruption to draw and/or describe in this arrow.

Think about how the disruption you chose impacts an ecosystem. Draw and describe 2 effects below:

©2021 Captivate Science

# INVASIVE SPECIES

MS-LS2-4

A species that is non-native or alien to an ecosystem. Introduction of invasive species causes harm to the environment, economy and/or humans.

Before completing this page, do some research about Zebra Mussels as an invasive species.

The Zebra Mussel is a small freshwater mussel native to Russia. It was accidentally introduced to the U.S. Great Lakes watershed via ballast water that was discharged from ships. Zebra mussels negatively disrupt ecosystems in many ways.

How does the Zebra Mussel invasion impact the Great Lakes ecosystem?

Answer these questions using specific EVIDENCE from the graphs provided.

**A** Invasive Mussel (Zebra) vs. Native Mussels (Unionid)

1. Based on graph A, when do you think the zebra mussel was accidentally introduced to the ecosystem?

1. Based on graph A, What impact has the zebra mussel had on the native mussel population? Why do you think this happened?

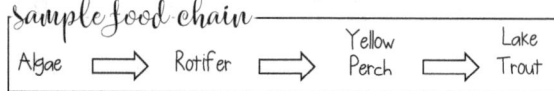

**B** sample food chain: Algae → Rotifer → Yellow Perch → Lake Trout

**C** Invasive Mussel (Zebra) and Rotifers in the Great Lakes

1. Based on the data in graph C, what can be said about the relationship between the rotifer population and the zebra mussel invasion? What can you infer based on this data?

1. Using data from A, B and C: What consequences might the Zebra mussel have on the Great Lakes food web? Give examples.

5. Using the evidence from data (A, B, C) and your scientific reasoning make a claim about how Zebra Mussels have impacted the Great Lakes.

Claim | Evidence | Reasoning

© 2021 Captivate Science

MS LS2-5

# BIODIVERSITY

The **variety** of species found in Earth's land and water ecosystems.

An ecosystem's biodiversity is often used as a measure of health.

## What can we do?

There are many ways to protect biodiversity. Read the solutions below and RATE them. Which do you think is most important? Rate the most important # 1.

_____ Learn as much as you can about nature and share your knowledge with others.
_____ Vote for government officials that want to help conserve ecosystems.
_____ Leave native plants undisturbed and landscape using only native trees and plants.
_____ Keep people and vehicles only on trails and main roads to reduce disturbance to wildlife and the spreading of invasive plants and animals.
_____ Conserve water and reduce irrigation to help maintain natural wetland areas.
_____ Use more natural products (for example natural soaps) to reduce pollution that leaks from sewer systems into freshwater and ocean ecosystems.
_____ Recycle to decrease pollution, energy consumption and the need for more garbage landfills.
_____ Avoid SINGLE use plastics (like straws). They are a major pollution issue in water ecosystems.
_____ Reduce meat consumption which impacts water and land biodiversity as well as creates harmful gases.

SHARE your ranking with a classmate. Do you agree or disagree and why?

# ECOSYSTEM SERVICES

MS LS2-5

The benefits people gain from ecosystems.

- Raw materials
- Food
- Medicine
- Recreation and Tourism
- Nutrient Cycling
- Soil Formation
- Erosion Regulation
- Water Regulation
- Climate Regulation

Label the wheel by matching each ecosystem service to the correct doodle picture.

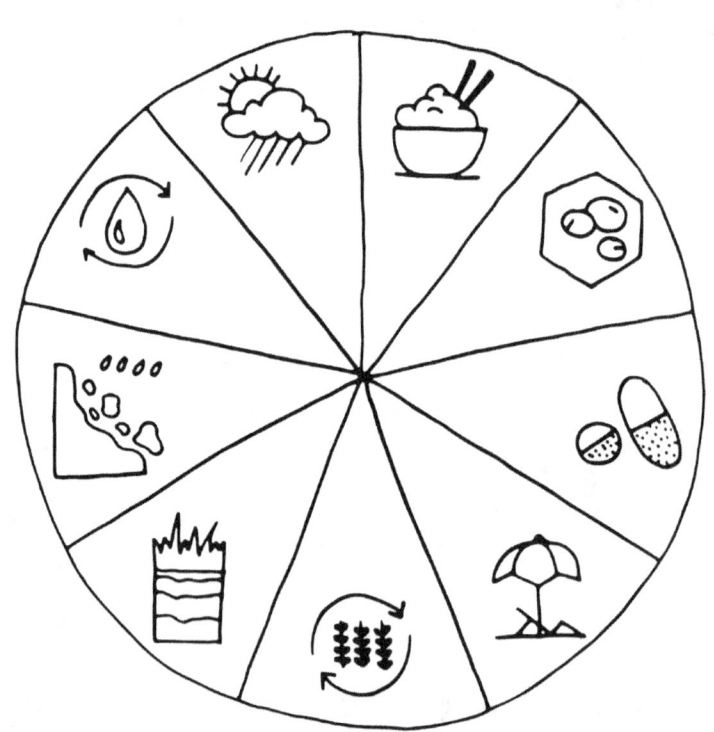

Describe your favorite ecosystem:

How can you tell if it is HEALTHY?

In what ways can humans have a positive impact on the ecosystem? Describe and doodle some solutions in the three circles below:

# CHAPTER 3

## Heredity: Inheritance and Variation of Traits

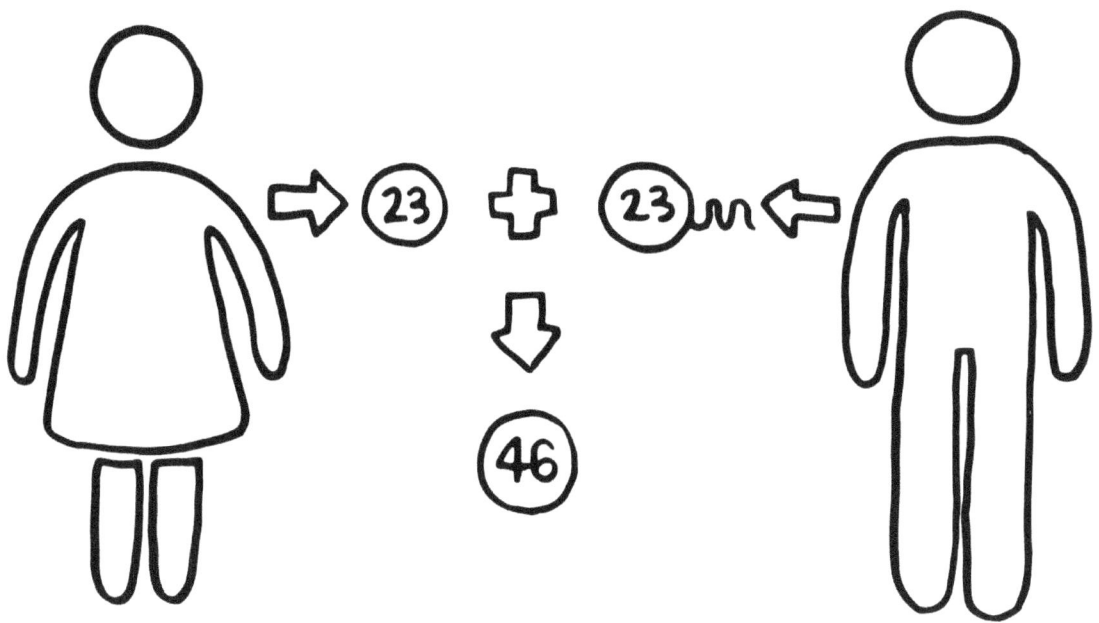

*How do living organisms pass traits from one generation to the next?*

*How do living things pass traits from one generation to the next?* **MS LS3-1**

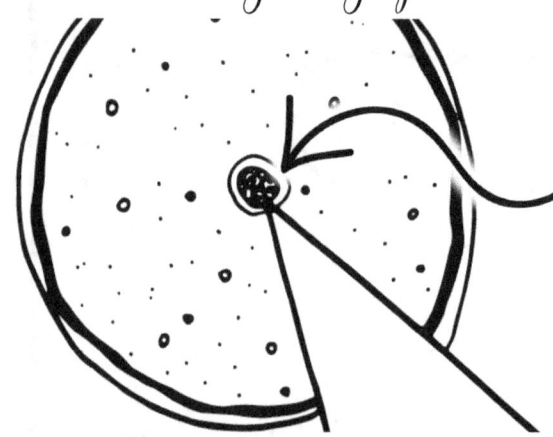

## Cell Nucleus

Did you know? Humans have 23 pairs of chromosomes, for a total of 46 chromosomes.

Thread-like structure made of coiled DNA found in the cell nucleus, that contains many genes.

Molecule that genes and chromosomes are made of.

**CHROMOSOME**

**DNA**

<sup>True</sup> <sup>False</sup>
**T** or **F** ?

1. Genes are located in Chromosomes. _____
2. Genes contain instructions for making proteins. _____
3. Genes are made of many Chromosomes. _____
4. Proteins affect the traits of an organism. _____
5. Chromosomes are found in the cell membrane. _____

In the space below, write a question you have about traits and heredity:

A unit of DNA that that contains instructions for making a protein molecule.

**GENE**

**PROTEIN**

A large molecule that depending on its structure and function, determines and organism's traits.

MS LS3-1

## INHERITANCE

Passing of genes from one generation to the next.

How many chromosomes do humans inherit from each parent? _____

Does everyone in your family have the same traits? Make a list of traits for each member of your house.

## VARIATION

Any difference in the traits between individual organisms.

All traits are determined by the environment or experiences of an organism. _____
All traits are determined by genes inherited from an organism's parents. _____

**AGREE or DISAGREE?**

EXPLAIN _____

## MUTATIONS

DNA Copying Errors. These are changes to genes that can result in in changes to protein molecules, which can affect the traits of an individual.

Gene mutations can cause variations that are...

Describe the three types of gene mutations in the boxes below. Use the graphics to help you!

# ASEXUAL REPRODUCTION

Reproduction that requires only ONE parent. The parent clones itself and offspring are genetically identical to the parent.

# SEXUAL REPRODUCTION

Reproduction that requires fertilization and results in offspring that are genetically different from either parent.

**VS.**

Name three organisms that asexually reproduce:

Name three organisms that sexually reproduce:

Describe one benefit and one drawback of asexual reproduction:

Describe one benefit and one drawback of sexual reproduction:

Describe the genetic composition of asexual offspring:

Describe the genetic composition of sexual offspring:

MS-LS3-2

**BIG IDEA** → In sexual reproduction, each parent randomly contributes half of the genes acquired by the offspring. Each cell contains two of each chromosome and therefore two alleles of each gene. These alleles can be the same or different.

**OFFSPRING** - An organism produced as a result of reproduction.

**ALLELE** - Variations of genes from each parent that result in different traits. Alleles are a specific form of a gene that provides instructions for making protein molecules.

**HETEROZYGOUS** - When gene versions from each parent are DIFFERENT

**HOMOZYGOUS** - When gene versions from each parent are the SAME

AA
aa

## WHAT IS THE DIFFERENCE BETWEEN DOMINANT AND RECESSIVE ALLELES?

A **dominant** allele produces a **dominant** trait in individuals who have _____ copy of the allele, which can come from just _____ parent.

For a **recessive** allele to produce a **recessive** trait, the individual must have _____ copies, one from each _____.

MS LS3-2

## Cause — DETERMINES THE → Effect

**GENOTYPE**: Genetic combination of alleles from parents.

And

**PHENOTYPE**: Physical expression or characteristic.

Brown is dominant, blonde is recessive. Parent one has brown hair and is heterozygous. Parent two has blonde hair and is homozygous. Can you predict the phenotypes of their offspring?

Parent one — Alleles: B  b

Parent two — Alleles: b  b

What percent of the offspring will have brown vs. blonde hair? _____

# MS LS3-1 AND LS3-2

## CAUSE and EFFECT

Use the **vocabulary** from standard MS LS3-1 or MS LS3-2 to write and/or doodle two different **cause and effect** statements that explain how living organisms pass traits from one generation to the next.

Try using some of these terms:

- Gene
- Protein Molecule
- Inherit (inheritance)
- Traits
- Mutation
- Variation
- Sexual Reproduction
- Asexual Reproduction
- Offspring
- Chromosomes
- Genotype
- Phenotype

You may also use other related words and phrases!

**If** _____

**then**

**If** _____

**then**

# CHAPTER 4

## Biological Evolution: Unity and Diversity

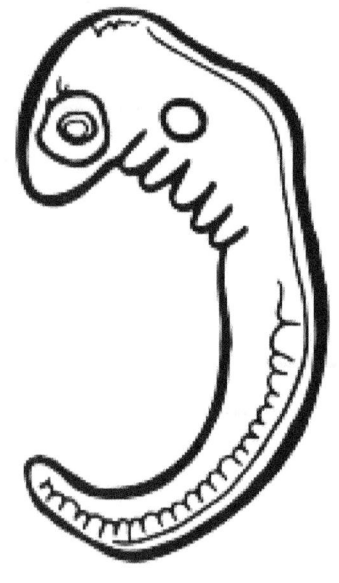 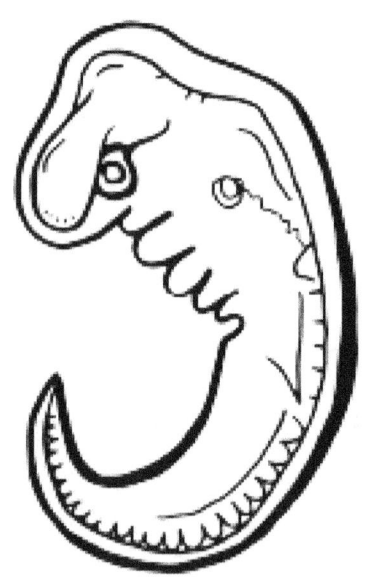

*How do organisms change over time in response to changes in the environment?*

# FOSSIL RECORD

The SUM of all the fossils or evidence of living things on Earth.

What can scientists learn by studying the fossil record?

_____
_____
_____
_____
_____

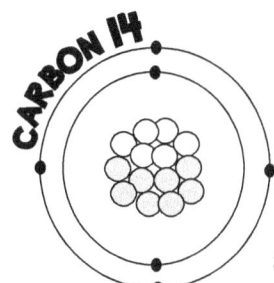
CARBON 14

# RADIOACTIVE DATING

A measurement of the amount of radioactive material that an object contains.

Carbon 14 breaks down at a known rate, so the amount of C-14 left reveals the age of the substance.

# RELATIVE DATING

Dating of events that places rock layers, geologic events and fossils in chronological order.

YOUNGER ↑ OLDER

Summarize and Re-state the information on this page to show you understand! Use writing and doodles that give ALL the details.

MS-LS4-1

Earth formed about 4.5 billion years ago. Life on Earth emerged about 2.1 billion years ago. Fossils and radioactive dating helps us understand the history of life on Earth.

MS LS4-1

## Billions of Years Ago

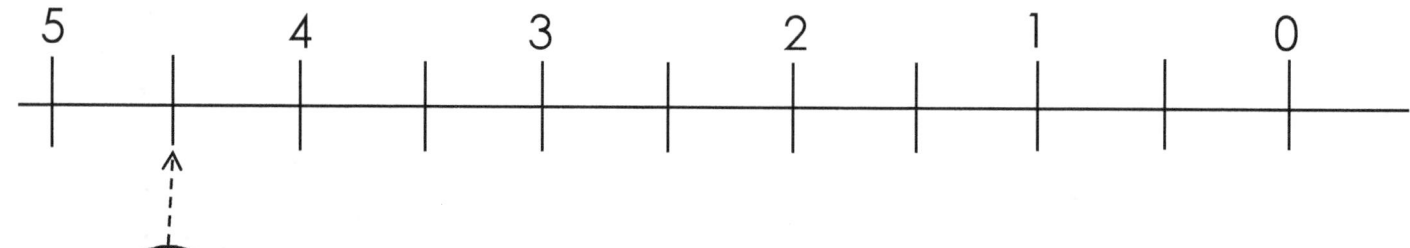

Formation of Earth

Add each life event below to the timeline above. The formation of Earth has been done for you. (You can also do some research and add some other events if you are a super scientist!)

First life on Earth (bacteria)
3.5 Billion Years ago

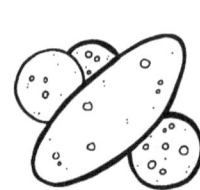
Protists
2.1 Billion Years ago

Plants and Fungi
1 Billion Years ago

Animals
600 Million Years ago

Some Animals Moved to Land
360 Million Years ago

Human Ancestors
7 Million Years ago

Approximately how long **after** the formation of Earth do the first living organisms appear in the fossil record? _____

Approximately how long **before** our first human ancestors did some animals move to the land? _____

Dinosaurs first appeared between 247 and 240 million years go. Between which two events above did Dinosaurs live?
_____

© 2021 Captivate Science

MS-LS4-1

When an organism no longer exists anywhere on the planet.

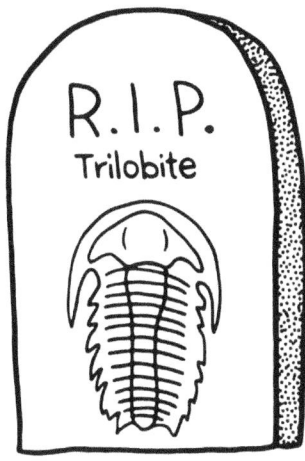

Name five or more organisms that are extinct. _____
_____

What is a <u>mass</u> extinction? _____
_____

Name some organisms that are currently listed as endangered _____
_____

How do you think the fossil record helps scientists learn about animal extinction? _____
_____
_____
_____

What impact does extinction have on biodiversity? _____
_____

Research: Did any animals go extinct recently? If so, write them here _____
_____

# What evidence shows that different species are related?

MS LS4-2

## ANATOMICAL SIMILARITIES

relating to body structure

Organisms with similar anatomical features are assumed to be related and they are assumed to share a common ancestor.

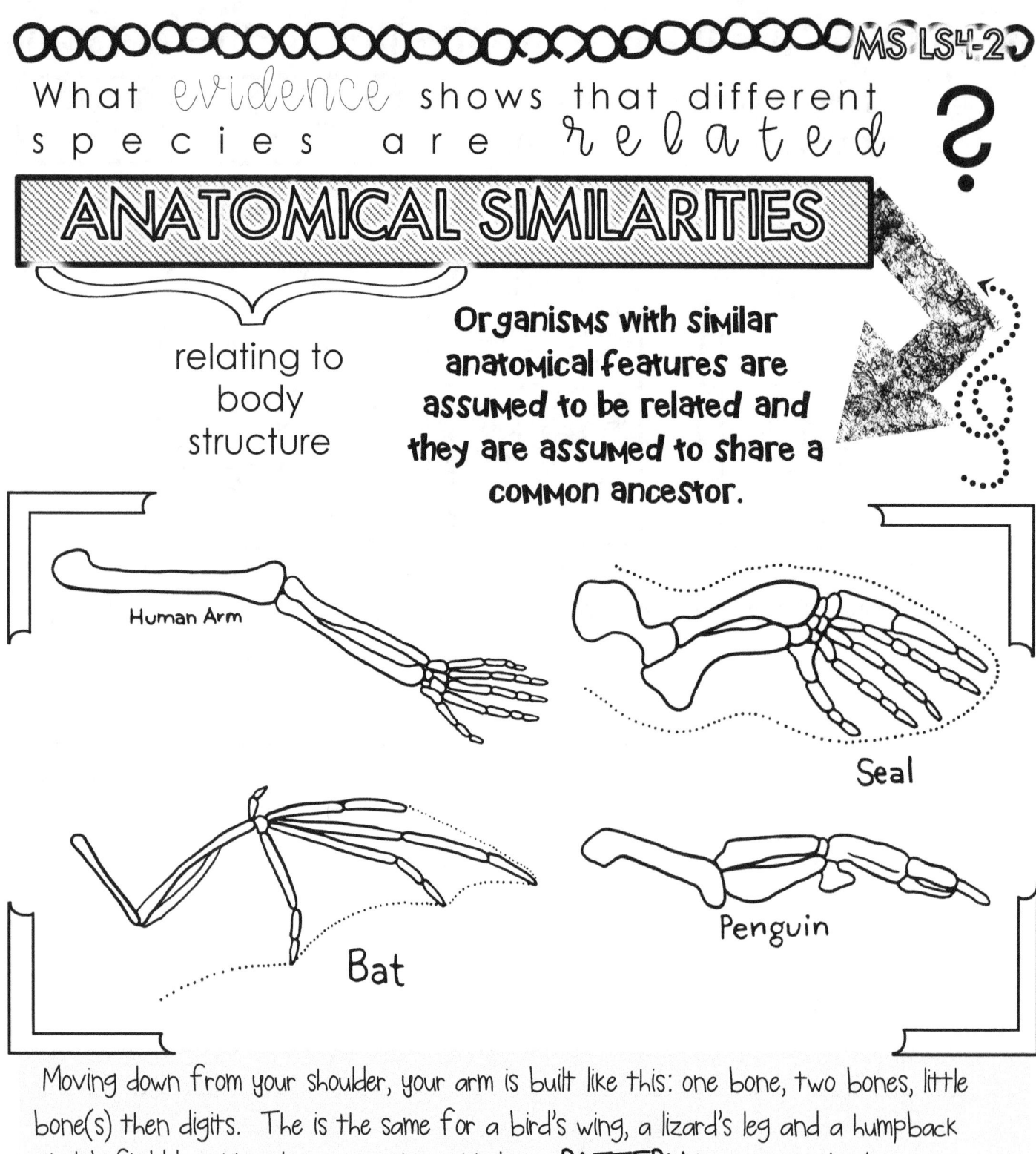

Human Arm

Seal

Bat

Penguin

Moving down from your shoulder, your arm is built like this: one bone, two bones, little bone(s) then digits. The is the same for a bird's wing, a lizard's leg and a humpback whale's fin! Use this color key to show this bone **PATTERN** in the animals above.

PURPLE- ONE BONE

GREEN- TWO BONES

YELLOW- LITTLE BONE(S)

RED- DIGITS (ex: fingers)

# EVOLUTION

MS-LS4-2

The change in characterisitics of species over many generations due to processes such as natural selection and mutation.

**RESEARCH** the evolution of a whale.

Record your source of info!

Website(s) used for information: _____
_____

Draw a **SEQUENCE picture** that shows how the animal has changed over time. Be sure to add LABELS to show important information about features and time.

| | | | |
|---|---|---|---|
| | | | |

## SAME
What are the SIMILARITIES between the modern animal and the animal of the past?

## DIFFERENT
What are the DIFFERENCES between the modern animal and the animal of the past?

Write a cool fact you discovered while researching!

## WOW!
_____
_____
_____
_____

MS.LS4-3

 The early developemental stage of an animal, while in the egg or the uterus of the mother.

 Are you related to fish?

What similarities do you notice that are not noticable in fully formed fish and humans? Add arrows, labels and/or color to identify similarities between the fish and human embryo.

### FISH EMBRYO
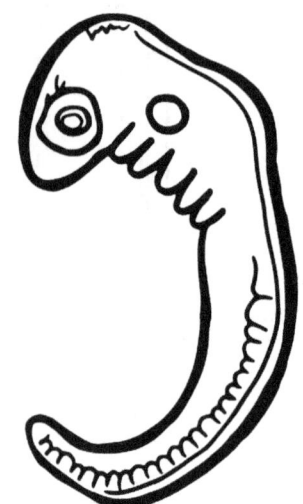

### HUMAN EMBRYO

What OTHER species have similarities to humans as embryos but not as fully formed anatomy? Research, write and doodle three more examples.

# NATURAL SELECTION

MS-LS4-4

The mechanism for evolution, where organisms with traits that favor survival, can produce more offspring, passing favorable traits on to the next generation.

Natural selection leads to **predominance** of certain traits in a population, and the **suppression** of others.

## SUPPRESSION

MUTATION CREATES VARIATION

To put an end to the activity of a person, place or thing. (Suppress)

In this visual model, which three organisms had traits that did NOT favor survival?

Shade in the circles to show your answer.

## PREDOMINANCE

The state of being greater in number. (Predominant)

In this visual model, which two organisms had traits that DID favor survival and were able to reproduce more?

Shade in the circles to show your answer.

MS·LS4-5

 Intentional reproduction of animals with desirable traits.

For Example

Process by which people breed animals or plants to develop particular physical traits.

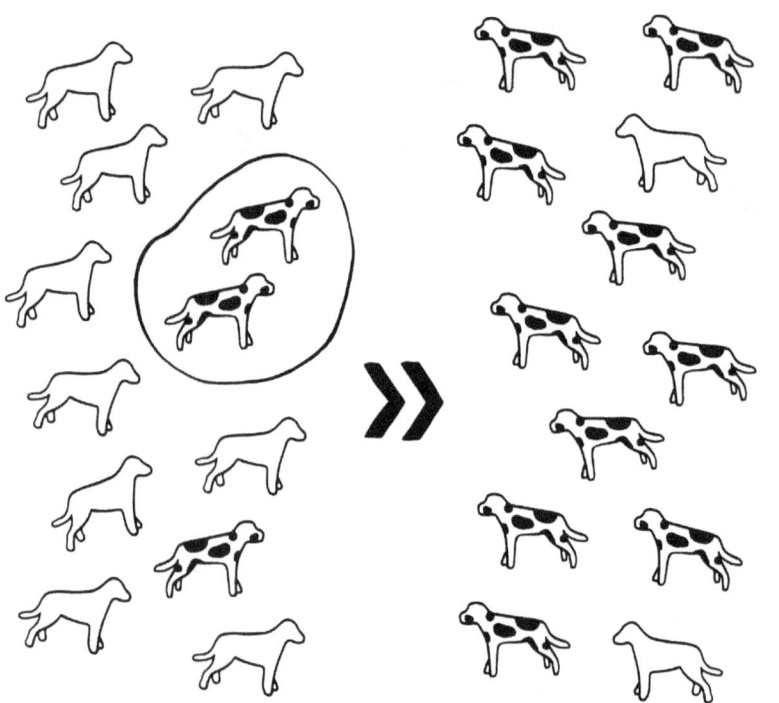

**THINK AND DISCUSS**

How does *genetic variation* among organisms affect survival and reproduction?

What are the pros and cons of artificial selection?

— MS·LS4-6

**THINK AND DISCUSS**

How does *environment* influence populations of organisms over multiple generations?

Natural selection leads to

**ADAPTATION**

A change that allows an organism to be better suited to its environment.

## More or Less?

Traits that support survival become _____ common.

Traits that DO NOT support survival become _____ common.

<u>Physical / Structural Adaptations</u> are changes in body structure or function that improve survival.

<u>Behavioral Adaptations</u> are changes in the way an animal acts/behaves in order to survive.

DOODLE and DESCRIBE 3 EXAMPLES OF EACH

| Physical / Structural | Behavioral |
|---|---|
|  |  |

# ANTIBIOTIC RESISTANCE

MS-LS4-6

Mutations create variation in bacteria resulting in different levels of resistance to antibiotics. Natural selection occurs when only the most resistant bacteria survive and reproduce.

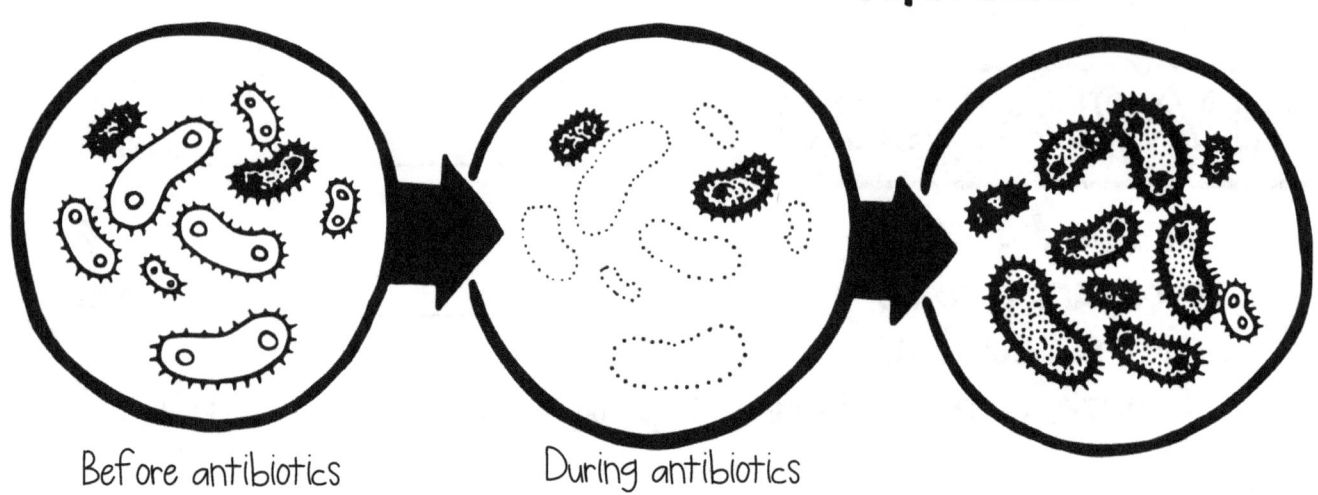

Before antibiotics | During antibiotics

Which bacteria were LEAST resistant to the antibiotics and died off? Draw one here:

Which bacteria were MOST resistant to the antibiotics and reproduced? Draw one here:

## DOODLE and DESCRIBE

DDT is a chemical that is used to control mosquito populations and help reduce malaria. In places where the chemical DDT has been used to kill mosquitos in the past, many of the mosquitoes are now resistant. Doodle and describe why this has happened based on what you know about **adaptation**.

# DOODLE DIY

## Doodle Templates for Addition Notes

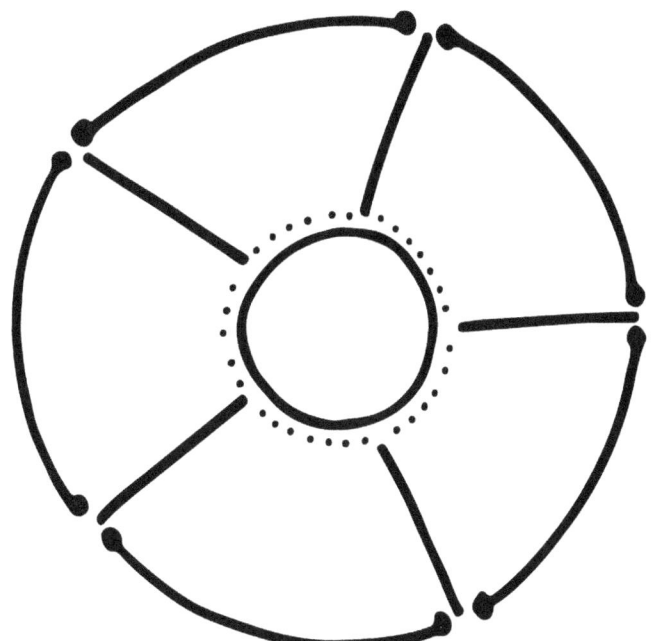

*These can be used for taking other notes to support your learning.*

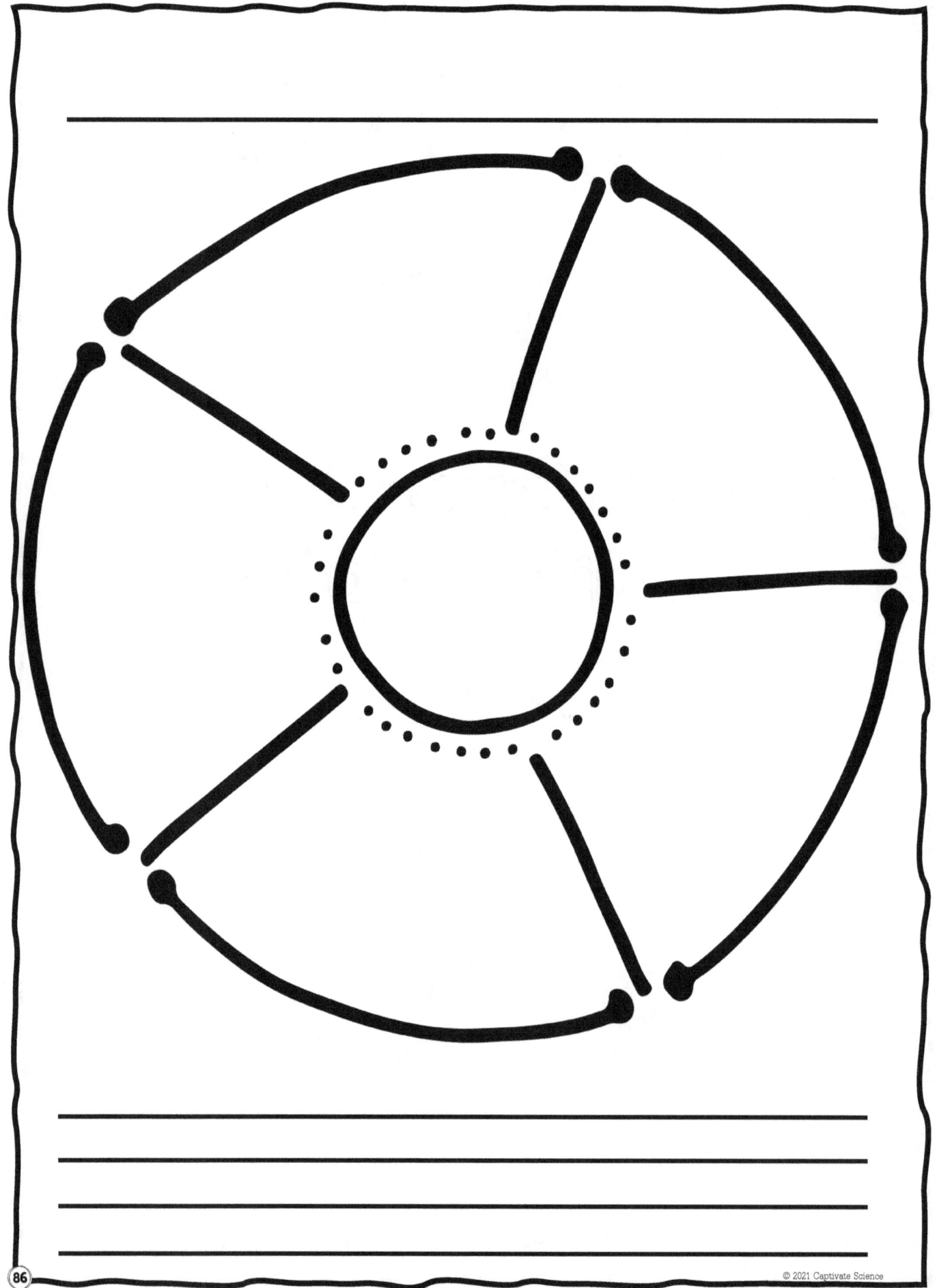

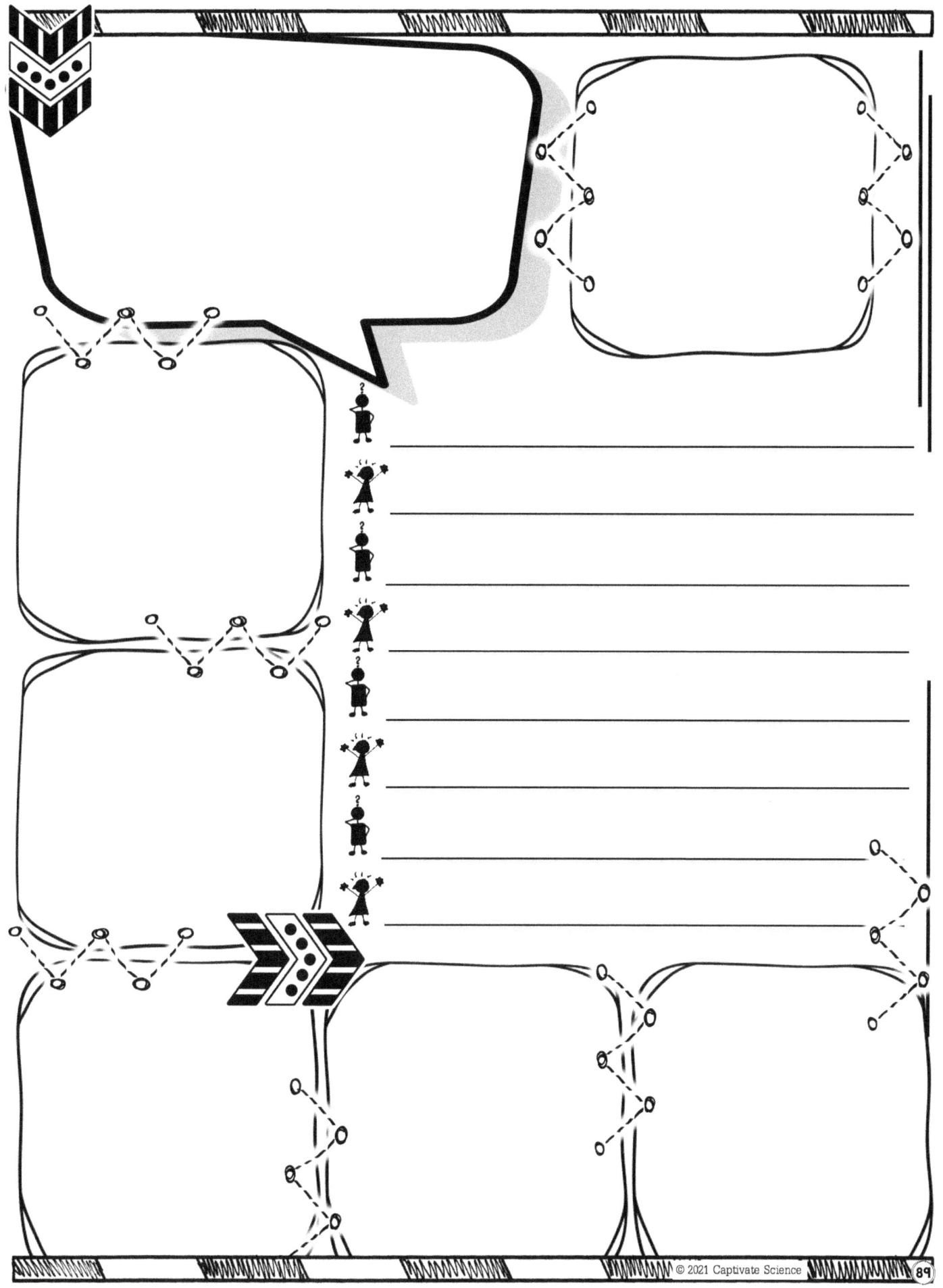

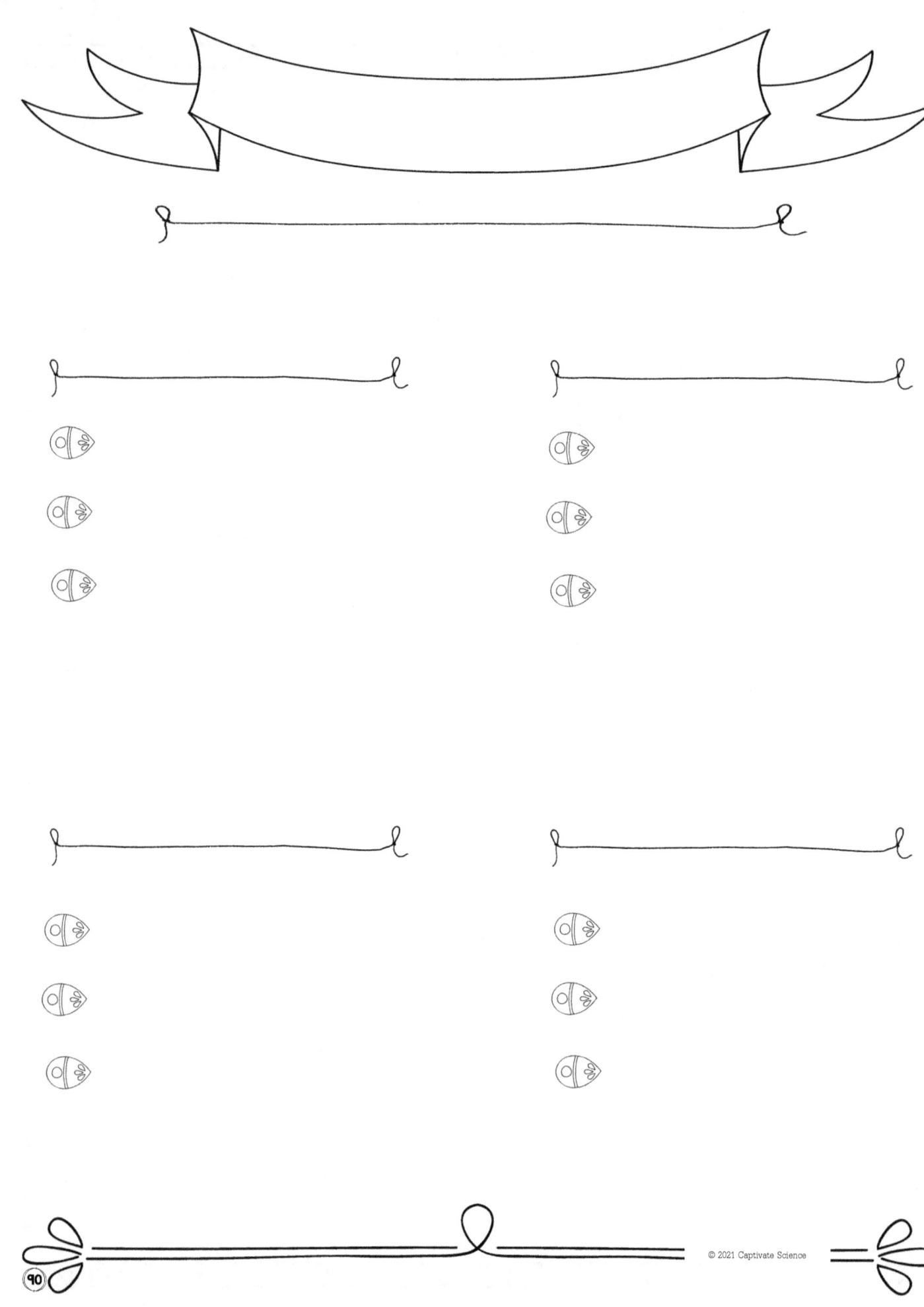

FULL COLOR
ANSWER PAGES

{OPTIONAL}
INTERACTIVE LINK PAGES

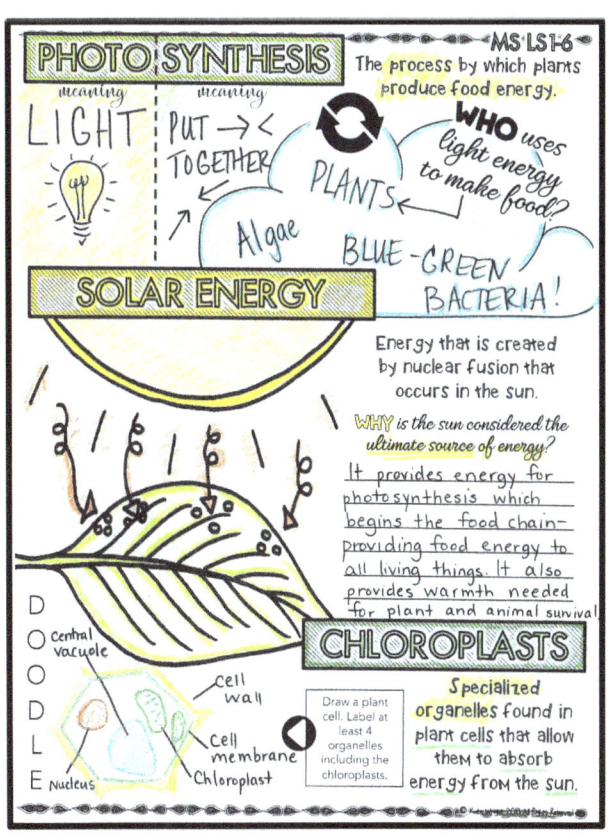

# Interactive Links and Full-Color Answer Guides for Each Page Available at captivatescience.com

PASSWORD: science

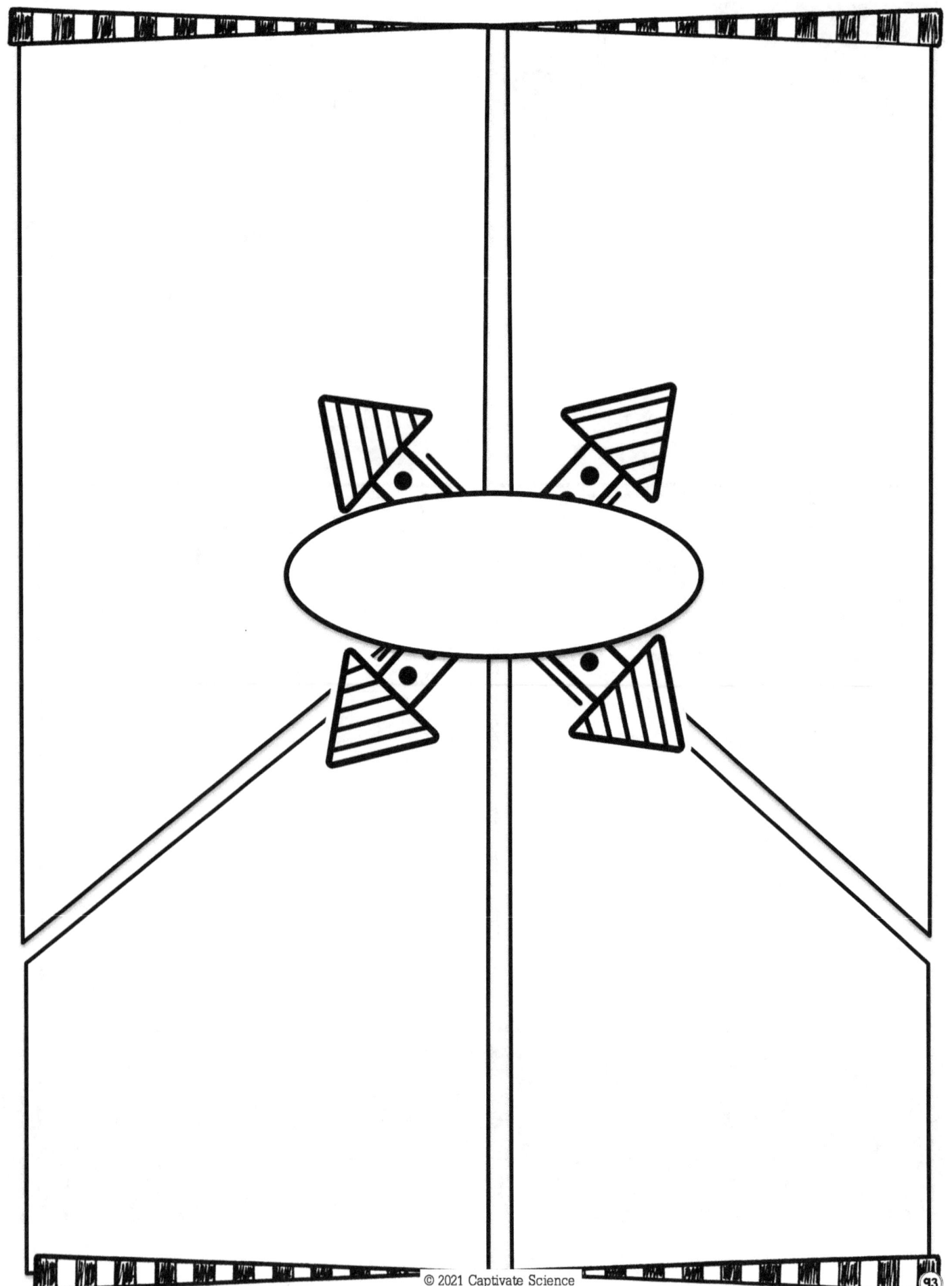

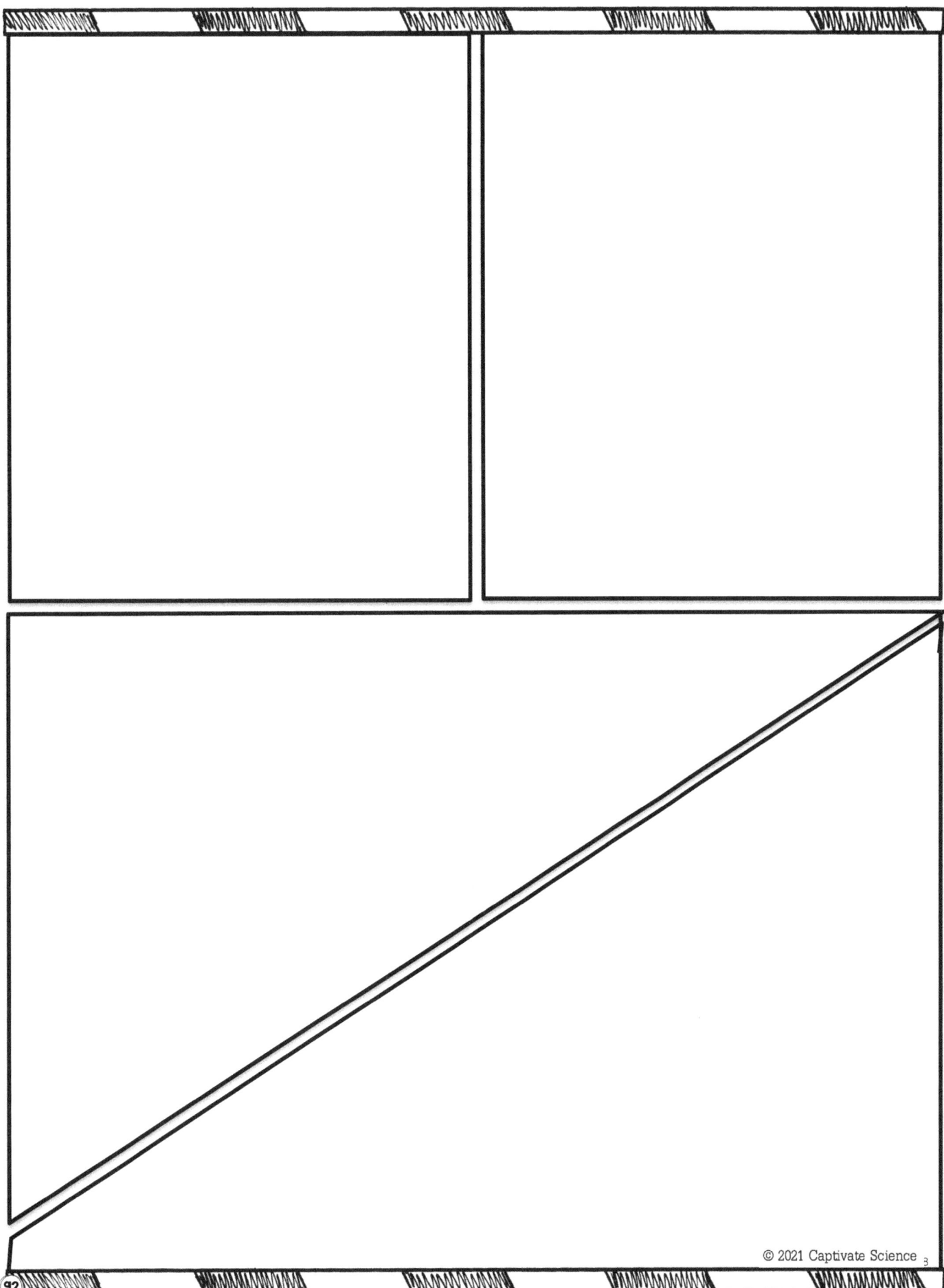

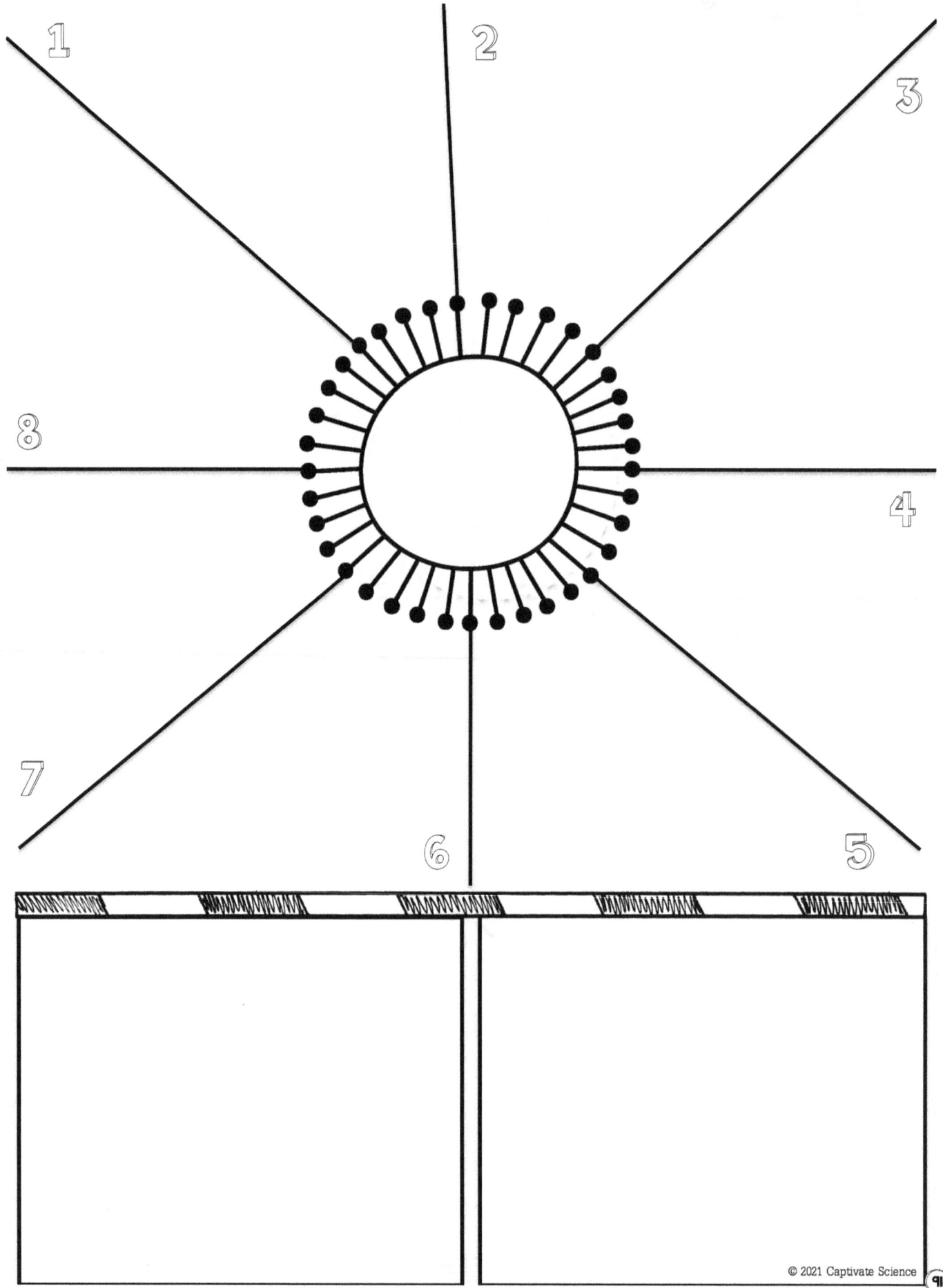

www.ingramcontent.com/pod-product-compliance
Lightning Source LLC
Chambersburg PA
CBHW081509080526
44589CB00017B/2705